57 Advances in Polymer Science

Fortschritte der Hochpolymeren-Forschung

Polymers in Medicine

Editor: Karel Dušek

With Contributions by
J. Drobník, R. Duncan, A. S. Hoffman,
Y. Ikada, J. Kopeček, F. Rypáček

With 52 Figures and 23 Tables

Springer-Verlag
Berlin Heidelberg GmbH
1984

ISBN 978-3-662-15245-4   ISBN 978-3-540-38740-4 (eBook)
DOI 10.1007/978-3-540-38740-4
Library of Congress Catalog Card Number 61-642

© Springer-Verlag Berlin Heidelberg 1984
Originally published by Springer-Verlag Berlin Heidelberg New York Tokyo in 1984
Softcover reprint of the hardcover 1st edition 1984

Typesetting: Th. Müntzer, GDR;

2152/3020–543210

Editors

Editorial

With the publication of Vol. 51, the editors and the publisher would like to take this opportunity to thank authors and readers for their collaboration and their efforts to meet the scientific requirements of this series. We appreciate our authors concern for the progress of Polymer Science and we also welcome the advice and critical comments of our readers.

With the publication of Vol. 51 we should also like to refer to editorial policy: *this series publishes invited, critical review articles of new developments in all areas of Polymer Science in English (authors may naturally also include works of their own).* The responsible editor, that means the editor who has invited the article, discusses the scope of the review with the author on the basis of a tentative outline which the author is asked to provide. Author and editor are responsible for the scientific quality of the contribution; the editor's name appears at the end of it.

Manuscripts must be submitted, in content, language and form satisfactory, to Springer-Verlag. Figures and formulas should be reproducible. To meet readers' wishes, the publisher adds to each volume a "volume index" which approximately characterizes the content.

Editors and publisher make all efforts to publish the manuscripts as rapidly as possible, i.e., at the maximum, six months after the submission of an accepted paper. This means that contributions from diverse areas of Polymer Science must occasionally be united in one volume. In such cases a "volume index" cannot meet all expectations, but will nevertheless provide more information than a mere volume number.

From Vol. 51 on, each volume contains a subject index.

Editors   Publisher

Table of Contents

Soluble Synthetic Polymers in Biological Systems

Jaroslav Drobník and František Rypáček

Institute of Macromolecular Chemistry, Czechoslovak Academy of Sciences, 16206 Prague 6, Czechoslovakia

This review summarizes facts and theories on the fate of a soluble polymer in an animal or human body, mainly from the point of view of therapeutic or diagnostic applications in medicine. Analytical methods which are the main source of our knowledge on this subject are discussed. A compartmental model is suggested and the importance of the individual compartmental barriers and the peculiarity of the polymer pharmacokinetics is documented: the molecular weight distribution changes with time in the circulating fraction; the various routes of parenteral administration are not equivalent. The main route of clearance from the body is the glomerular filtration, which is limited by the size and modified by the charge of the molecule. Polymers which pass through glomerulus may be reabsorbed in the tubulus by pinocytosis, which is largely dependent on the chemical nature of the polymer. Synthetic polymers may elicit antibody production, induce immunological tolerance, activate supressor cells, serve as a non-specific immunostimulator, particularity of the macrophage cytotoxicity, etc. Practical applications in medicine require more detailed knowledge of the fate of polymers in the body. Diagnostics is more a promising field for the immediate future.

Advances in Polymer Science 57
© Springer-Verlag Berlin Heidelberg 1984

Abbreviations

DIVEMA	regular 1:2 copolymer of maleic anhydride and divinyl ether
DNP	2,4-dinitrophenol
GPC	gel permeation chromatography
HES	hydroxyethylstarch
MTX	methothrexate
PEG	polyoxirane (poly(ethylene glycol))
PHEA	poly-α,β[N-(2-hydroxyethyl)-D,L-aspartamide]
PHPMA	poly[N-(2-hydroxypropyl)methacrylamide]
PVA	poly(vinyl alcohol)
PVP	polyvinylpyrrolidone
PVPNO	poly(vinylpyridine N-oxide)
RES	reticuloendothelial system

1 Introduction

All living matter consists primarily of polymers. Therefore, when products or parts of organisms are used for any purpose, polymers of biological origin are used. In addition to applications in food, clothing, construction, etc., biological polymers have been used for healing purposes. Although nature provides a huge selection of different polymers, it is sometimes difficult to select a natural macromolecule which would fulfil all demands of pharmacy and medicine. Recent progress in macromolecular chemistry had led to the hope that synthetic polymers could be tailored to fit in desired functions in the body better than natural polymers. While the use of natural polymers has mostly been based on empirical experience, synthesis for a special purpose requires complete theoretical understanding of the role the synthetic polymer would play in the body. Our ignorance in this respect is the main barrier to the introduction of synthetic polymers in medicine.

Any substance of natural or artificial origin entering the *milieu interieur* of the organism must be considered as a "foreign body". Its interactions with the components of the biological environment determine the promptness of recognition and the intensity of the reaction of the organism in eliminating or isolating the intruder and re-establishing the internal equilibrium. From this point of view, it is quite correct to judge the application of synthetic polymers in the organism very carefully as the introduction of a foreign substance. Similar to other artificial invaders, including many synthetic drugs as well as the surgeon's scalpel or x-rays employed by the examining doctor, the ratio of positive and adverse effects is the decisive factor in the application of polymers in therapy and diagnostics. The understanding of their behaviour and fate in the organism must be sufficiently complete to allow the reliable evaluation of the benefits and risks. This problem is not, of course, the subject matter of a single discipline; polymer chemistry, physiology, pharmacology, immunology, medical science, biochemistry and other disciplines should combine in its solution.

In addition to the properties of a given polymer, the complex biological mechanisms involved in the handling of the polymer by the organism participate to a major degree in the fate of the polymer in the body. It is the purpose of this review to reveal the relationships between the particular polymer properties and the biological mechanisms they participate in and to indicate how many different factors, influences and rules must be considered when the fate of a synthetic soluble polymer in a living body is to be understood. In order to include at least the most important factors, we could not avoid describing some fundamental biological pathways and terms, that are already familiar to readers with a background in biomedical polymer research and that are, on the other hand, useful for explaning this subject to chemists who are just beginning to participate in this field.

2 Analytical Methods

Analytical methods are the only source of direct information on the fate of the polymer in the body. They include detection, identification and quantitative estimation of the polymer. Detection my be defined as tracing of the polymer by chemical techniques in the bulk of biological samples or as morphological localization in organs, tissues and cells by histological and cytochemical procedures. Identification of the polymer should distinguish the polymer from other macromolecules in the biological milieu and, at the same time, should reveal all changes, both physical and chemical, that may occur with the polymer during its history in the organism. Quantitative estimation is a complex problem, particularly when the polymer has changed. If it has been degraded, then all of its products (metabolites) should be detected. Some typical approaches will be discussed.

2.1 Radioisotopes as Tracers of Synthetic Polymers in the Body

Radioactive labelling is based on either incorporation of the radioisotope into the polymer structure or on attachment of a radioactive tag to the polymer. All labelling methods are well known and have frequently been described; however, they may also be a source of certain pitfalls. These questions will be discussed later.

The incorporation of a radioactive isotope into the structure of the polymer leads to practically no changes in the chemical and physical properties of the macromolecule. It also enables tracing of the degradation products from biodegradable polymers. This advantage is offset by the tedious preparation. This procedure can hardly be used for pharmacological evaluation of industrial products, as it is practically impossible to prepare a labelled laboratory sample identical in all respects with the industrial product. Thus, this method is limited to research applications. Artefacts can be generated by isotope exchange and the effects of their specific activity are discussed below.

The attachment of a radioactive tag to the polymer is a much more versatile method; however, inherent changes of the macromolecule may result. The significance of these alterations should be assessed from the viewpoint of the purpose of the experiment. A compromise between this and the former method involves the use of a radioactive polymerization initiator which is incorporated into the polymer chain. However, the specific activity which can be achieved is very low.

Iodination of the polymer with ^{125}I or ^{131}I is analogous to the method routinely used in biochemistry and results in only small changes in the polymer properties. However, the nature and the stability of the iodine-polymer bond must be carefully considered. For example, the use of radioactive iodine for labelling of PVP has been questioned [1]. The use of GPC or treatment with AgI powder in vitro and measurement of the radioactivity of the thyroid gland in vivo are believed to be sufficient control measures.

The attachment of a small molecule containing a radioisotope results in more alteration of the macromolecule; e.g., the acylation of amino groups by ^{14}C-acetic anhydride leads to changes in the charge and hydrophilicity reflected in the electrophoretic pattern [2]. We have studied the excretion of poly[(N-2-hydroxyethyl)-D,L-aspartamide]s (PHEA) of molecular weight about 10,000, where the 2-hydroxyethyl group was partly replaced by the 4-hydroxyphenetyl group which can be easily and safely iodized. We found that, up to 4 mol-% substitution, there were no changes in the excretion and deposition of the polymer in the kidneys. This does not imply that other biological effects are insensitive to the same level; e.g., immunological processes may be expected to respond at a much lower level of substitution [3].

The labelling with iodine radioisotopes showed the advantages of gamma emitters: they can be measured with minimum sample preparation, in a well-counter in whole organs and even in whole small animals, and with proper collimation in the body of patients and big animals by gamma cameras from outside. For patient's safety, short-lived nuclides are preferable. Because of the simple preparation in generators ^{113m}In and ^{99m}Tc are widely used in radiodiagnostics. Thus, labelling should be performed immediately before application and must, therefore, be limited to a very simple procedure. Attachment to the polymer can be achieved through a chelating group [5,6] which, however, by no means represents a negligible alteration of the polymer structure.

It follows from the labelling procedures discussed above that ^{3}H and ^{14}C nuclides are most often incorporated into the polymer molecule, as heteroatoms (S, P, etc.)

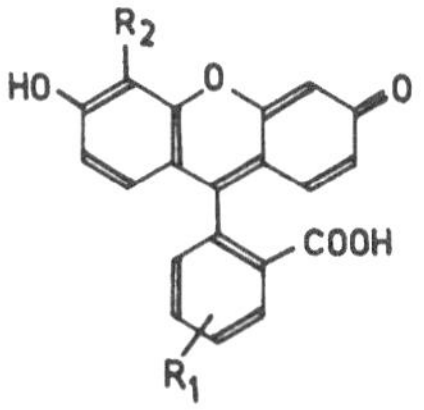

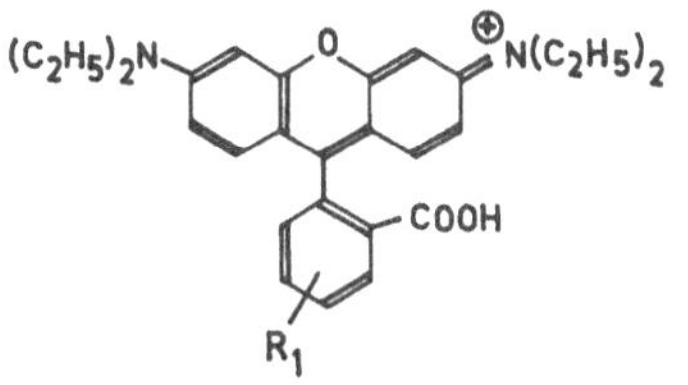

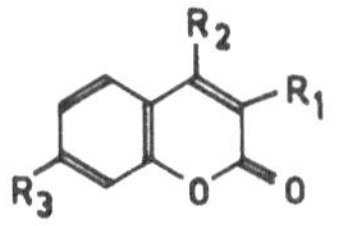

		R$_1$	R$_2$	R$_3$	Ref.
	a	-H	H	-	
	b	-NCO	H	-	15)
	c	-NCS	H	-	16)
I.	d	-NHCSNHCH$_2$CH$_2$NH$_2$	H	-	26)
	e	-NHCOCH$_2$CH$_2$NH$_2$	H	-	26)
	f	-NHCSNH(CH$_2$)$_2$CONHNH$_2$	H	-	27)
	g	-H	-NHCO-C(CH$_3$)=CH$_2$	-	11,12)
	a	-H	-	-	
II.	b	-NCS	-	-	16)
	c	-NHCSNH(CH$_2$)$_2$CONHNH$_2$	-	-	27)
	a	-H	H	H	
III.	b	4-C$_6$H$_4$-NCS	-CH$_3$	-N(C$_2$H$_5$)$_2$	17)
	c	-H	-CH$_2$NH$_2$	-OH	13)
	d	-H	-CH$_2$NHCOC(CH$_3$)=CH$_2$	-OH	13)

Fig. 1. Derivatives of fluorescein (Ia), rhodamine B (IIa) and coumarin (IIIa) that are most useful in fluorescence labelling

are rarely components of the polymer structure. The nuclide ^{14}C is the tracer of choice for bulk detection and estimation of the polymer and its possible metabolites. It is less suitable for morphological localization by microautoradiography because of its inherent low specific activity. It should also be noted that ^{14}C-labelled compound are quite expensive.

The use of ^3H avoids most of these problems. High specific activity can be cheaply obtained by catalytic exchange of protons in many compounds. The energy of the emitted electrons is low, yielding very sharp microautoradiograms. However, this simple labelling methods have some drawbacks: care must be taken to check reverse exchange with protons in vivo and during all laboratory manipulations. The availability of a high specific activity makes it particularly important to take radiochemical and/or radiobiological considerations into account when working with ^3H.

First, it should be noted that each decay is accompanied by a chemical transmutation: $^3H \rightarrow {}^3He$, $^{14}C \rightarrow {}^{14}N$, $^{32}P \rightarrow {}^{32}S$, $^{35}S \rightarrow {}^{35}Cl$. The daughter atom receives the recoil energy from the emitted electron and is in the electronically excited state [6]. Second, the emitted electron ionizes the atoms it passes by stripping off their valence electrons losing about 33 eV per ionization. All these events usually lead to bond rupture in the vicinity of the transmutating nuclide and to energy transfer to the surrounding atoms and molecules. The density of ionization increases with decreasing electron energy, i.e. velocity. For example, each disintegration of ^3H yields an amount of energy equal to 10 rads ($0.1\ J\ kg^{-1}$) to a sphere 1 μm in diameter [7]. Thus, energy transfer to the surroundings must be considered at high specific activities. It may generate chain scission, oxygen activation, radical generation and serious biological damage to the structures where the polymer accumulates. We have calculated [8] that the lysosomal membrane, in experiments with tritiated poly(acrylic acid)[9] with a specific activity about $10^{11}\ Bq\ g^{-1}$, received a dose of 12 krads ($120\ J\ kg^{-1}$) which, is more than sufficient to change its permeability [10] and thus generate artifacts.

In conclusion, radioactive labelling is a method of choice of a very sensitive tracing of the polymer and its possible metabolites in bulk biological material. It is less suitable for morphological analysis. Care must be taken to check for alternation of the polymer properties, the stability of the label and for radiochemical and radiobiological side-effects.

2.2 Fluorescent Labels

The labelling of polymers with fluorescence labels cannot avoid the addition of a new structure to that of the original unlabelled polymer. This obstacle need not be serious in the experimental research (in which labelled polymers are mostly used); however, the effect of perturbation of the properties of the polymer should be checked specifically in each experiment. If these precautions are taken, fluorescence labelling can be very useful.

In addition to minimal perturbation of the original polymer structure, the polymer-fluorochrome bond must be stable and the labelled polymer should have favorable spectral properties. The highly fluorescent derivatives of fluorescein (Ia), rhodamine (IIa) and coumarin (IIIa) have been studied in detail for analytical purposes in biological material (Fig. 1). Attachment of these substances to a polymer molecule by a covalent bond may be accomplished either by copolymerization of a polymerizable derivative of the fluorochrome or by a polymeranalogous reaction.

Polymerizable fluorescent vinyl monomers of fluorescein (Ig) [11,12] and coumarin (IIIa) were prepared [13]. Although, the polymerization of a well-defined fluorescent monomer seems to be an exact method for the preparation of well-defined labelled polymers, in practice this is rarely true. It is usually desirable to study polymers bearing not only the label but also other groups of special biological interest or useful for additional binding of such compounds. Therefore, at least terpolymerization should be the starting procedure. In addition, fluorochromes in general can easily form radicals, and may enter the polymerization in an unpredictable way either as a chain transfer or as a terminating group. On the other hand, these reactions were found to be useful for the preparation of fluorescent polymers, e.g. polyacrylamide, by polymerization in the presence of fluorescein as a chain transfer agent [14].

A polymeranalogous labelling reaction may employ either the reactive electrophilic group on the fluorochrome or on the polymer. For the former, the isocyanates and isothiocyanates, e.g., Ib [15], Ic, IIb [16], IIIb [17], are preferable and their usefulness has been confirmed in many experiments with labelled proteins (see Ref. 18 for a review). This approach was also followed in labelling of soluble polymers of biomedical interest, e.g. dextran [19,20] and other polysacharides [21,22], poly(ethylene oxide) [23], poly(ethyleneimine) [24], poly(2-hydroxypropyl methacrylamide) (PHPMA) and PHEA [25]. This method appears attractive because of its apparent simplicity, but our experience has shown that the results are not always satisfactory. The high reactivity of these derivatives may involve them in reactions with different types of nucleophilic groups on polymers. While the reaction with an amine yields a substituted urea derivative with sufficient hydrolytic stability at neutral and acid pH values, the product of reaction with a hydroxy group — present in most hydrophilic polymers — is substantially less resistant to hydrolysis [25]. Thus, labelling of hydrophilic polymers with fluorochrome isocyanates or isothiocyanates results in different types of polymer-fluorochrome bonds exhibiting also different degrees of stability.

It is usually not difficult to prepare polymers containing various reactive electrophilic groups or to activate polymers for labelling by the latter method. Sometimes this step is already included in the preparation of tailored polymers. The primary amino group is then a suitable nucleophilic group for the fluorochrome. Derivatives of fluorochromes with an aliphatic amine have been prepared (Id, Ie in Ref. 26 and IIIc in Ref. 13) and their reactions, hydrolytic stability of the bond formed with polymer as well as their spectroscopic properties have been studied [13,26]. The acylation of an aliphatic amine, which proceeds most readily, does not affect the emission properties of the fluorochrome. The best results were obtained with the label Ie having an amide bond. In another approach, hydrazides prepared from isothiocyanates of fluorescein and rhodamine (If and IIc) [27] were used. This method was originally suggested for polysaccharides, but may be extended to the labelling of other polymers.

While the radioisotopic labelling method described above has some unquestionable advantages, the fluorescence labelling method has its outstanding features as well: the sensitivity of quantitative assays is at least comparable with isotopic methods. An amount of about 10 ng of polymer tagged with one mole of fluorochrome per thousand moles of monomer units can be quantitatively determined in biological material using standard equipment [26], If more advanced systems for fluorometric assay are used, the threshold of assay is decreased to a fluorochrome (fluorescein) concentration of 5×10^{-14} mol l^{-1} [28]. This offers, for example, the possibility of simple and convenient flow-through GPC and molecular weight distribution analysis in tiny samples of biological fluids [29]. In addition to quantitative measurements, fluorescence spectroscopy, including microfluorimetry, may also yield structural information on the nature of the fluorochrome microenvironment [30], interactions with cell structures [31], the mobility of side chains on polymer, etc. The morphological localization at a histological and cellular level of polymer deposition can readily be achieved with a fluorescence-labelled polymer. Cytofluorographs can be used for counting and separation of cells containing the polymer [32].

On the other hand, the nonspecific fluorescence background of some compounds occurring in the living tissues and fluids (liver, kidneys, bile etc.), emission quenching, and sorption of the polymer on components of the studied material place rigid requirements on careful preparation of the sample for quantitative measurements at a high sensitivity level [26]. Comparison with labelling by γ-ray-emitting isotopes emphasizes this fact. Possible alterations of the emission properties due to chemical modification of the fluorochrome, particularly as a result of the action of detoxicant enzymatic systems in the living body (see Chapter 4), should be also considered. Acetylation of 3'- and 6'-hydroxy groups of fluorescein (essential for emission) in plant cells has been reported [33] as well as glucuronidization in rat liver cells [34]. It is not clear, however, whether the macromolecular substrate can undergo the same type of reaction. Therefore, fluorescence labelling need not be fully reliable for long-term tracing of a polymer in the body, when the absolute quantity of the polymer in the tissue is the most important factor. Nevertheless, such advantage as simple and safe handling, high sensitivity, easy visual morphological and flow-through detection, low price, no problems with waste products, etc. may outweigh the above-mentioned shortcomings in many experiments.

2.3 Other Methods

Several other methods of polymer quantification in biological material have been used. In general, a polymer is usually isolated from biological material by a deproteinization procedure, extraction with organic solvents, etc., and then analyzed. Elemental analysis [35], viscosimetry [36], turbidimetry [37] and complexation with iodine [38,39] (for PVP) in extracts were used in early studies in this field. These methods are now rarely used and have been replaced by labelling methods. Among the classical analytical techniques, the anthrone reaction for the estimation of carbohydrate polymers (e.g. dextran [40], inulin [41]) has remained useful.

The morphological detection of synthetic polymers on the tissue and cellular level either by microautoradiography, microfluorography, or other methods has some general features worth of a more detailed discussion. Historically, the first demonstrations of polymer deposits in tissue were based on the observation of morphological changes resulting from the presence of the polymer in cells without direct identification of polymer material. Bargmann was the first to describe the swelling and vacuolization of the spleen reticulocytes and Kupffer cells of liver after administration of PVP [42]; most subsequent histological studies dealing with the storage of soluble polymers in the body are based on these indirect observations. Several attempts have been made to stain the deposited polymer using histological stains [43-46], but these methods did not exhibit the desired specificity.

Autoradiography can directly identify and localize radiolabelled polymers in tissues [47]. Autoradiograms have very impressively shown the overall topography of polymer distribution in whole-body [48] or organ sections [49]. Similar techniques have been adopted for fluorescence-labelled polymers [50].

However, the growing interest in targeting drugs carried by polymers and, consequently, in the investigation of the mechanisms of polymer interactions with the cells surfaces and cellular capture requires more detailed morphological information. Both radioactive and fluorescence labelling could probably provide polymer identification, but the resulting accuracy in polymer localization mostly depends on the success in the preservation of its original in vivo deposition pattern. In vivo, polymers are stored in precisely localized deposits, because of the semipermeability of biological membranes (see Chapter 3). The procedures and chemicals commonly used in histology for the fixation of tissues impair the semipermeability of biological membranes but, unfortunately, they usually leave the synthetic polymers soluble (not only in water but also in organic solvents) and thus susceptible to diffusion. They may be washed out or adsorbed artificially on other structures. The dispersion of a polymer in a cell may also occur during long-term exposure in the autoradiography of frozen sections kept at a temperature slightly below zero (-5 to -20 °C) [51], since these temperatures are not low enough to solidify the concentrated electrolytes in frozen cells. However, these obstacles can be overcome using appropriate methods. Rapid procedures of tissue fixation using glutaraldehyde and osmium tetroxide have been developed in electron microscopy for ultrastructural localization of dextran as well as PVP deposits [52,53]. The immediate *post mortem* perfusion of the blood system of the organ under study by fixation solution containing paraformaldehyde and glutaraldehyde was successfully used in preserving the true cellular localization of fluorescence-labelled polyaspartamide polymers [50] (Figs. 9, 10). A method based on ultrarapid deep-freezing of the tissue sample, followed by controlled freeze-drying and subsequent vacuum embedding in a paraffin-like medium was suggested as generally applicable for any water-soluble polymer [54]. This procedure decreases to a minimum the possibility of artificial translocation of polymer during the preparation of the sample.

In conclusion, most analytical problems of tracing the polymers in the body can be solved by proper labelling. Radioisotopes are very helpful in bulk analysis, whereas fluorescent tags provide excellent morphological information. Of course, modification of the polymer structure in labelling must be considered. In morphological studies by any method, the crucial role is played by the sample preparation. All steps from killing the animal to mounting on the slide should be carefully checked to prevent the movement and delocalization of the polymer.

3 Movement of Polymers in the Body Compartments

3.1 The Compartment Model of an Organism

Polymers entering the *milieu interieur* of an organism cannot move randomly but their movement is controlled by anatomical and physiological barriers. In pharmacokinetics, which describes and quantifies the dynamic processes of absorption of chemicals in the body, their distribution to various tissues, reactions with tissue components and their elimination from the tissue and the body *via* metabolism and/or excretion, it has appeared useful to represent the body by a system of compartments. These compartments need not have physiological or anatomical counterparts. While the body is composed of an infinite number of compartments, in pharmacokinetics, a compartment refers to all those organs, tissues and cells for which the rates of uptake and subsequent clearance of a chemical are sufficiently similar to preclude pharmacological resolution [55].

We will utilize such a multicompartment model of an organism to describe factors and mechanisms controlling the movement in the body of a macromolecular substrate — a synthetic water-soluble polymer. Compartments in our definition are characterized rather by similarity in mechanisms of crossing of compartmental barriers. The macromolecular character of polymers is an important factor in these crossings. Biopolymers, which are endogenic in nature, i.e., mainly proteins, play an indispensible role in the homeostasis of the organism, both structural as well as functional, and at the same time they all participate through their macromolecular nature in the determination of the osmotic, rheologic and ionic properties of body fluids. An exo-

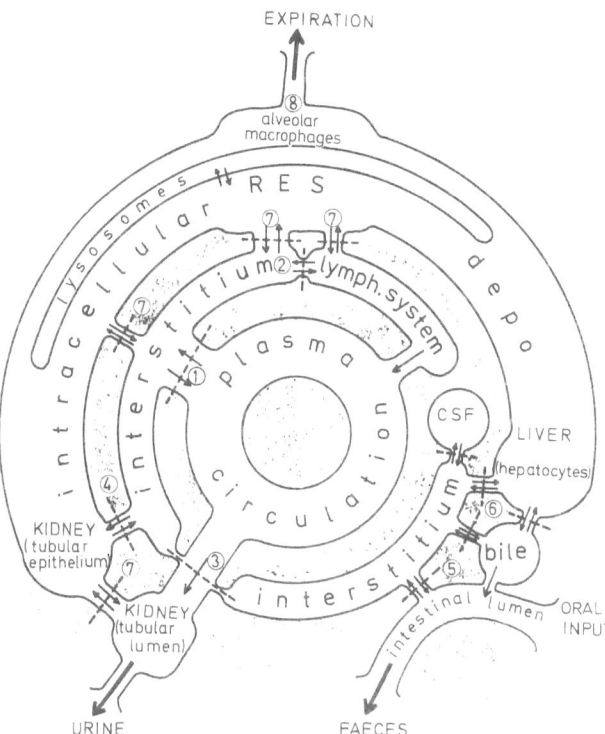

Fig. 2. Multicompartment model of organism. Areas depict the body compartments, connecting corridors represent possibilities of polymer transfer between compartments either restricted by compartmental barriers (dashed lines) or occurring as flux transfer (simple arrows). The numbers refer to the paragraphs in which the given barrier crossing is discussed

genic macromolecule, e.g., a synthetic polymer, entering any compartment will disturb its equilibrium state, and therefore may be either transferred to other compartments or subjected to processes which usually tend towards re-establishment of the original equilibrium.

In general, during its transfer the foreign macromolecule may "employ" the transport mechanisms "already prepared" for endogenic macromolecules. Obvious exceptions are due to the fact that polymers under consideration are mostly unable to fit into the biologically specific mechanisms developed during the evolution for particular biomacromolecules. The limited biodegradability of synthetic polymers leads to the most serious consequence of this general feature.

In Fig. 2, the body compartments are schematically depicted as areas connected by corridors symbolizing the pathways of possible solute exchange between compartments. The pathways are interrupted by dashed lines (barriers); the mechanisms of crossing these barriers are the subject of the following paragraphs. Simple arrows indicate unrestricted flux transfer for solutes. The central compartment — plasma circulation — is the only compartment through which the exchange of compounds between remote parts of the body may be accomplished. The large intracellular compartment, joined graphically in one area, comprises all the cells of the body. The participation of cells of the reticuloendothelial system (RES), kidney tubular epithelium, and liver hepatoxytes is so important that it should be discussed separately.

3.2 Ways in which Synthetic Polymers Cross Compartmental Barriers

The distribution of a polymer in the body, the rate of its clearance, the site and time duration of its retention, i.e., all the basic factors determining the availability of a medical polymer for biological activity may be understood as resulting from its partitioning at compartmental barriers. These particular barriers may have various compositions (see below) but mostly their crossing includes some means of crossing of a biological membrane, which will be briefly characterized below.

A biological membrane (a plasma membrane, plasmalemma, cell membrane, etc.) is a lipoprotein membrane (Fig. 3). Lipids are arranged as a double layer of molecules with the hydrophobic portion facing inwards and hydrophilic portion outwards, and are maintained in this position by hydrophobic forces. Between the lipid molecules, the proteins are intercalated, being bound to the lipid layer by hydrophobic interactions. The proteins are not held rigidly in place but can move laterally along the lipid membrane forming a fluid mosaic pattern. Many cell membrane proteins have polysaccharides bound to their outer surfaces, they may participate in determining the antigenic specificity, and together with others secreted by the cells, they form a coating layer, called the glycocalyx. The glycocalyx may be of various thickness and density on various cells. In principal, this structure of the lipoprotein membrane (unit membrane) is common for all other membranes enveloping the cell organels (nucleus, lysosomes, plasmalemma vesicles, etc.) [172].

The transport of low-molecular weight compounds (water, ions) is not accompanied by any visible morphological changes. It proceeds by passive diffusion or an active transport process, depending on the polarity and size of the transported molecule and presence of a specific carrier protein in the membrane. Water-soluble polymers

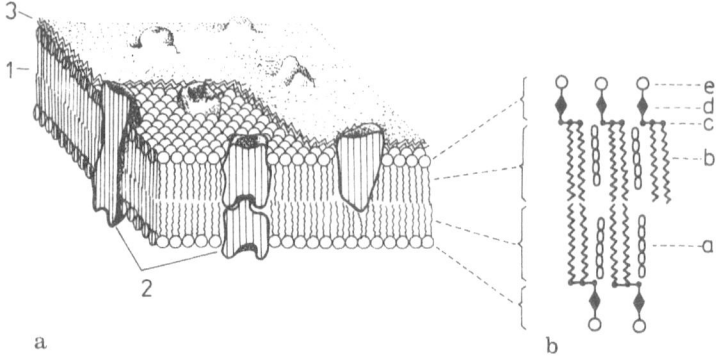

Fig. 3a and b. Fluid mosaic structure of the cell membrane. **a** Overall representation: 1 — bilayer of lipid molecules; 2 — membrane proteins, including receptors, membrane transport proteins, proteins mediating signal transfer etc.: 3 — polysaccharide layer — glycocalyx; **b** molecular structure of the lipid bilayer: a — cholesterol; b — lateral chains of fatty acids; c — glycerol residue; d — phosphate moiety of the phospholipid molecule; e — positively charged (amino) part of the phospholipid molecule (From Refs. [63,172])

cannot pass across the lipoprotein membrane by any kind of diffusion process without interuption of the integrity of the membrane. The occurrence of defects in the lipoprotein membrane has been reported resulting from a perturbation of the cell surface, by the electrostatic field, exerted by polyelectrolytes [56] or high concentrations of poly(ethylene glycol)s [57]. It was supposed that a small fraction of macromolecules may find their way into the interior of cells through such openings [58]. These phenomena have been studied in vitro either with artificial membranes or with isolated cells in the absence of a serum which compensates these effects [59]. Thus, the internalization of macromolecules through cell membrane defects is doubtful in vivo.

A common way in which macromolecules can cross a biological membrane includes the invagination of the membrane, the formation of a vesicle which then buds inwards from the membrane and loses contact with it. During this process, the macromolecule enters the cell completely enclosed in a membrane vesicle, together with a certain volume of extracellular fluid. This process, generally termed endocytosis, includes ingestion of fluid material (pinocytosis) and ingestion of particles (phagocytosis [60]). Vesicle transport seems to be the only way in which synthetic polymer can overcome the barrier formed by the lipoprotein membrane [61,62].

3.2.1 Transendothelial Passage of Polymers

A polymer introduced into the blood stream (central compartment) circulates in a closed system of blood vessels. It can be cleared from this compartment either by the endocytic activity of specialized cells (see cells of RES) or by passage through the walls of the blood vessels. Transport of solutes across the vessel wall attains great importance at the microvasculature level (arterioles, capillaries, venules). Capillary walls are formed almost solely by the endothelium, a monolayer of cells, highly flattened, roled up to form a tube (Fig. 4) [63]. The cells are held together by tight junctions, structures in which glycocalyx layers of two adjacent cells are fused. In most capillaries, these junctions are generaly impermeable to molecules larger than 1.8 to 2.0 nm in diameter [64] whereas the endothelium of the venous part of vasculature — postcapillary venules — has junctions opened for molecules approx. 6.0 nm in diameter [65]. Only in some tissues, e.g. in the liver, spleen, bone marrow and the lymph nodes, are the capillaries of the sinusoidal type, i.e. without basal membrane and with endothelial cells not tightly joined, leaving sufficient intercellular space for free communication with the interstitial fluid [63] (a milieu surrounding all tissue cells). With these exceptions, the larger molecules could probably cross the endothelial wall mainly by passing through the cell volume, i.e. crossing two (inner and outer) cell membranes. This process involves transport of plasmalemmal vesicles which are formed by pinocytosis-like action on one side of the endothelium and which traverse toward the opposite side of the cell and discharge their content by exocytosis [60,65]. The vesicles are about 70 nm in diameter and their transport proceeds in both directions [66]. In addition, the fusion of several vesicles together or simultaneously on both sides of a highly flattened cell may occur. This is thought to be the mechanism leading to the formation of temporary transendothelial channels [65] (Fig. 4). This vesicle transport may be equivalent to the large pores predicted by physiologists from measurements of rates of transport of macromolecular tracers between blood and lymph [67-69]. The flux transfer of solutes through the gaps in intercellular junctions and through transendothelial channels is forced by arterial hydrostatic pressure, while the vesicle transport is more probably diffusion controlled [70,71]. Several types of macromolecular tracers have been used including proteins [68,72], polysaccharides [73,74] and synthetic polymers, mainly PVP [75-79], in studies on endothelial permeability for macromolecules.

The density of plasmalemmal vesicles as well as stability of the transendothelial channels varies in different tissues, and is accompanied by a variation of the permeability of the vascular endothelium for polymers, which is, in general, inversely de-

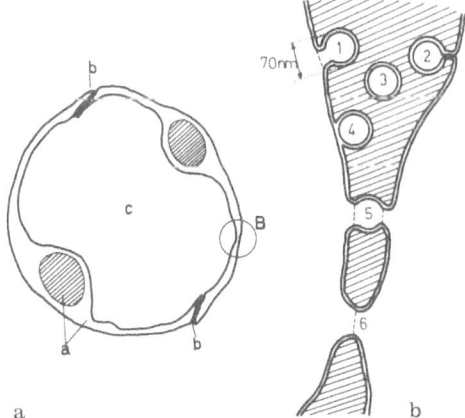

Fig. 4a and b. Diagram of the capillary endothelium structure **a** with schematic depiction of various stages in the vesicle transport **b. a** a — endothelial cell, highly flattened with nucleus; b — intercellular tight junction; c — capillary lumen; **b** 1 — formation of pinocytic vesicle; 2 — detachment of the vesicle from the membrane; 3 — vesicle transfer through the cytoplasma; 4 — fusion-fission of the vesicle with the membrane; 5 — simultaneous fusing of the vesicle with both membranes leading to the formation of a transendothelial channel or the fenestrae (6) (from Refs. [63,65])

pendent on the molecular size. The permeability is greatly diminished in the vascular system of brain. Brain capillaries are formed by endothelium cells which not only lack in pores and rarely contain plasmalemmal vesicles but also have junctions that are impermeable for macromolecules [80]. Because of this "blood-brain barrier", synthetic polymers are not usually found in the brain tissue [58,81]. If they are introduced directly into the cerebrospinal fluid, they are rapidly cleared by pinocytosis of specialized cells [82].

3.2.2 Transport into the Lymphatic System

Liquid passing through the capillary wall constitutes the interstitial fluid. Macromolecules may be drained from this compartment either back into the venous capillaries with pores in the intercellular junctions larger than the arterial pores or through the lymphatic system. Lymphatic vessels originate in tissues as blind capillaries. They are formed of a single layer of endothelial cells with large openings in the intercellular junctions [63]. Their permeability to large macromolecules is generally much higher than that of the blood capillaries. The composition of the lymph — the fluid drained by the lymphatic vessels — is therefore very similar to the composition of the interstitial fluid [75,83,84]. Thus, the polymers leaving the central compartment and entering the interstitial compartment, are differentiated according to their size, the larger molecules being drained preferentially *via* the lymphatic system [85,86]. The lymphatic vessels gradually converge in two large trunks which empty into the large veins near the heart — returning the macromolecules back to the central compartment. Interaction of the polymer with RES and immunocompetent cells may take place during passage through the lymphatic system.

3.2.3 Glomerular Filtration

The endothelial permeability of glomerular capillaries of the kidneys occupies one of the central positions in the handling of synthetic polymers by mammalian organisms. Glomerular capillaries are fenestrated by many open pores. The interstitial space is reduced (see Fig. 5c) and cells of the visceral epithelium adhere to capillaries due to many finger-like processes which are separated from one another by narrow spaces [63]. The capillary endothelium and the processes of visceral tubular epithelium are separated by a continuous basal membrane which has a fibrous network composition. Because of the presence of substantial amounts of mucopolysaccharides and glycoproteins containing sialic acid, the basal membrane has an overall negative charge [87,88]. The pores in the endothelial cells together with the sieving properties of the basal membrane are the basic features of the glomerular filter. In addition to the structural characteristics of the glomerular filter, the passage of macromolecules across the walls of glomerular capillaries is also determined by glomerular hemodynamics [89-91] and molecular parameters of macromolecules [92-97].

During the normal operation of mammalian kidneys, a large amount of fluid (about one fifth to one third of the plasma volume) passes through the endothelium of the glomerular capillaries. This process may be regarded as ultrafiltration, resulting from an imbalance in transcapillary hydraulic (ΔP) and osmotic pressures ($\Delta \pi$). The flow rate J_v can be written as

$$J_v = k(\Delta P - \Delta \pi) = k[(P_{GC} - P_T) - (\pi_{GC} - \pi_T)]$$

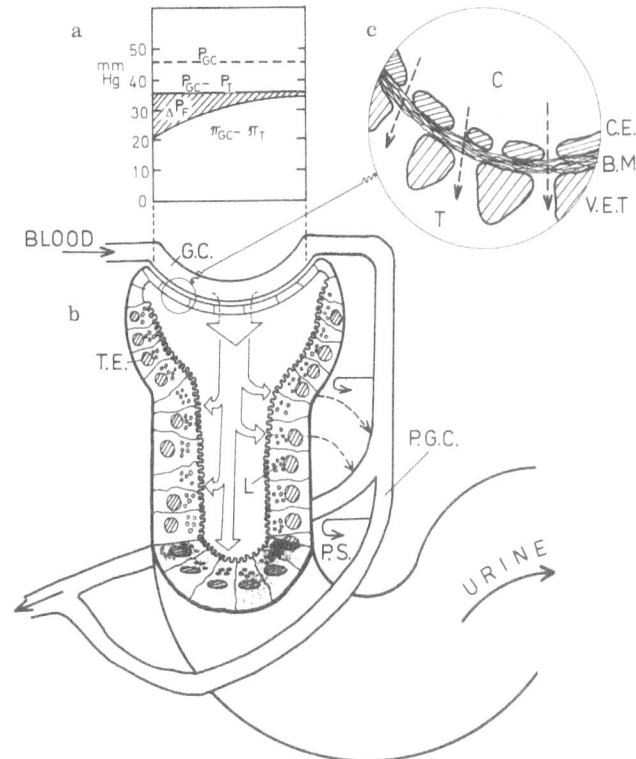

Fig. 5a–c. Schematic depiction of structural and functional relationships in the transport of polymers in the nefron — a basic functional unit of the kidneys.
a A diagram of glomerular filtration pressure: P_{GC} — hydrostatic pressure of the blood in the glomerular capillary; P_T — hydrostatic pressure of liquid inside the tubulus; π_{GC}, π_T — coloid osmotic pressure in the capillary and tubulus respectively, the hatched area (ΔP_F) represents the filtration pressure along the glomerular capillary (Ref. [90]); **b** Schematic depiction of the glomerulus and proximal part of the tubulus: liquid filtered from the glomerular capillaries (G.C.) is mostly reabsorbed by the cells of the tubular epithelium (T.E.) — open arrows. The low molecular weight compounds together with water are transported by the epithelial cells in the peritubular space (P.S.) and absorbed in the postglomerular capillaries (P.G.C.) — dashed arrows. Macromolecules are pinocytosed in the epithelial cells and are either degraded in the secondary lysosomes (L) or stored there if they are nondegradable. Transport of polymers from postglomerular capillaries into urine (tubular secretion) is hindered by the tubular epithelium — reversed arrows. **c** Enlarged view of the structure of the glomerular barrier: C — lumen of glomerular capillary; T — lumen of tubulus; C.E. — fenestraeted endothelium of capillary wall; B.M. — basal membrane; V.E.T. — finger-like processes of cells of visceral epithelium of tubulus. Transport from the glomerular capillary into the lumen of the tubulus is not hindered by the tubular epithelium

where the subscripts GC and T denote the pressure in the glomerular capillary and in the tubular lumen, respectively [90] (see Fig. 5). The model of the glomerular filtration was drawn on the basis of several studies involving various dextran fractions [93,95,96,98], as well as PVP [78,99,100] and PEG [96,101] and using GPC for the measurement of the size distribution of polymers in the urine and plasma. In this model, the permeability properties of mammalian glomeruli were described in terms of an isoporous membrane. It has been suggested that the pore radius is about 5.0 nm [97]. The fractional clearance of macromolecules decreases with increasing molecular radius (Fig. 6) but, for a given size, the transport of polyanions is selectively restricted and filtration of polycations is enhanced [102]. The effect of molecular charge is usually connected with the presence of the fixed negative charges in the pores of the glomerular membranes [97]. In addition, the fractional clearances of proteins — which are macromolecules with a compact structure — are generally much lower than those of flexible polymers (PVP, dextran, PEG) of a given molecular radius. Therefore, several authors have inferred that other parameters such as shape, flexibility and deformability of the polymer coil may also play an important role in the transport of polymers across the glomerular barrier [89]. It is thought that the transport of linear flexible polymers may be facilitated by "end-on" movement [95]. This explanation is based on experimental studies of the transport of linear polymers through the gels and concentrated solutions of hyaluronic acid [105, 106].

Deformation of the molecule of a flexible polymer might occur as a consequence of the pressure gradient existing during glomerular filtration (see Fig. 5A). It can be calculated that this shear stress for the assumed parameters of the pores in the glomerular membrane may attain a value about 26 N/m². It is doubtful if this value is sufficient to produce appreciable flow deformation of the polymer coil [96]. Jörgensen and Møller [96] compared the glomerular clearances of narrow fractions of dextran and PEG and found that the glomerular permeabilities for these two polymers are inversely proportional to their degree of expansion in aqueous solutions. Linear PEG, forming a more expanded structure in water than partially branched dextran, was also more restricted by the glomerular membrane (see Fig. 6). The relationship between the glomerular filtrability and the size follows the same pattern with both

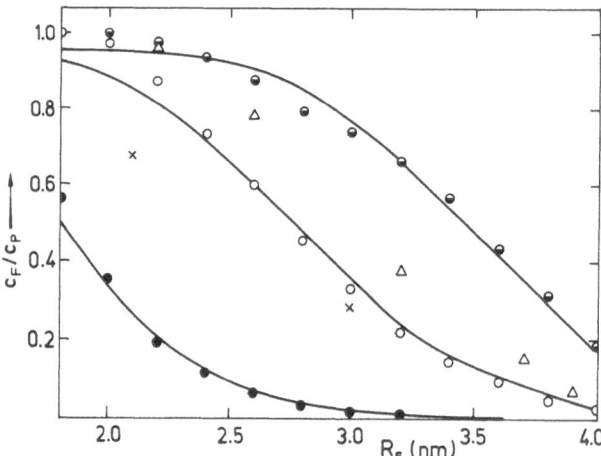

Fig. 6. Glomerular filtration of polymers. Filtrate to plasma concentration ratio (C_F/C_P) as a function of molecular size for neutral and charged polymers (R_s — Stokes radius). Points represent average experimental values obtained in rats for neutral dextran (○ ref. [97] △, Ref. [96]), dextran sulphate (● ref. [97]) diethylaminoethyldextran (◐ Ref. [97]) and polyethylene glycol (× Ref. [96]). The solid curves were calculated using a model assuming a pore radius in the glomerular membrane of $r_o = 4.7$ nm and a charge density in the membrane of 165 meq/l (Ref. [97])

polymers if the polymer size is expressed in terms of the effective radii, obtained from the GPC data. These findings have thrown doubt on the above mentioned assumptions, as they did not indicate any perturbation of the effective molecular size by shear stress deformation.

3.2.4 Tubular Secretion

Because of the peculiar arrangment of the epithelial cells in glomeruli, polymers are filtered from the central compartment directly into the urinary system through which they can be excreted out from the body. Other ways of excretion of polymers include crossings of epithelial barriers. One of these ways might be represented by a tubular secretion. It has been supposed [35,47,107,108] that polymers larger than the molecular weight limit for rapid glomerular filtration (about 25,000 for PVP) could still be excreted *via* the kidneys through pores in the postglomerular capillaries (i.e. capillaries in the peritubular space (see Fig. 5) the permeability of which is higher. It has been suggested that these capillaries have pores permeable to PVP with molecular weights as high as 650,000. Polymers can pass through these pores into the interstitial fluid of the peritubular space and can be drained from here either by the venous or the lymphatic capillaries [78]. However, the above mentioned theory does not explain how the polymers pass across the barrier represented by the tubular epithelium (see Fig. 5b). While a number of low-molecular-weight compounds can be transported from the peritubular space into the urine through the epithelial cells of tubuli, i.e. *via* the tubular secretion, there is no evidence that a polymer can be transported intact either from the peritubular (contraluminal) to the luminal side of the epithelium or in the opposite direction. No evidence supporting such assumption can be drawn from the fate of other macromolecules [109]. Although pinocytosis from the contraluminal side of tubulus was reported for some peptide hormones, e.g. insulin [110,111], it proceeds at a much lower rate than pinocytosis from the luminal side [109]. In addition, this pinocytosis seems to exhibit a high degree of specificity, distinguishing even between the homologous and the heterologous insulin [112]. It seems also sufficiently proved that macromolecules cannot pass between the epithelial cells constituting the tubular walls in normal kidneys [109].

The question of whether polymers pass between the cells of the epithelia or only through them by vesicle transport is also relevant in the discussion of polymer passage into the intestine lumen and into the bile. Both routes are often assumed in the literature as ways in which polymers can leave the body.

3.2.5 Intestinal Transport

The data on the relative role of the vesicle transport and diffusion through the intercellular junctions located between the epithelial cells are scarce. The excretion of PVP into the intestine has been described several times. The available information can be briefly summarized as follows: The fraction of the (i.v. administered) PVP excreted in the faeces of rats was about 4% of the administered dose after the first day [113], about 7% after three days [108,113] and about 13% after eight days [108]. Excretion through the intestine may continue even after kidney excretion falls below the detection limit [113]. The fraction excreted by the intestinal pathway is reported to be less dependent on the molecular weight than the kidney excretion [114]. This relative independence of the molecular size may indicate participation of the vesicle transport mechanism, but no attempts have been made to estimate the molecular weight distribution of the excreted polymer and, therefore, no direct support for this suggestion is available. Nevertheless, it is believed that intestinal excretion may be a way in which large molecules, that cannot be degraded or excreted through the kidney, can excape from the body.

Transport in the reverse direction, i.e. intestinal absorption of polymers after oral administration, has been demonstrated to depend on the molecular weight [115,116]. Large molecules are mostly unable to undergo intestinal absorption [117]. In a study with ^{125}I-labelled PVP, its absorption was found to be independent of the amount of ^{125}I-PVP infused, suggesting the existence of a readily saturated absorption mechanism [118]. The presence of a mucus layer on the luminal side of the epithelial cells may contribute to a selective exclusion of macromolecules (see below) [119].

3.2.6 Biliary Transport Route

A fraction of the intestinaly excreted polymer may be transported into the intestinal lumen with the bile. Bile is produced by the liver parenchymal cells (hepatocytes) by active secretion in the sense that the hepatocytes transform and transport blood components, bile acids and bilirubin into the bile canaliculi. There are some indications that macromolecules and even particles, e.g. colloidal mercuric sulfite, colloidal Ag, can be eliminated from blood in the bile [120,121], although this elimination is slow. De Duve points to the presence in rat bile of significant amounts of all lysosomal hydrolases, as support for the suggestion that the particles and macromolecules may appear in the bile as a consequence of emptying of some secondary lysosomes by extrusion [122]. In this way, even polymers larger than the threshold of glomerular filtration could be exocytosed into the bile and excreted *via* the intestine out of the body, if they are formerly pinocytosed by the hepatocytes (Fig. 7).

Excretion in the bile was followed with PVP [108] and polyaspartamides [29]. The results of both studies are in good agreement in the sense that they both revealed preferential excretion of the low molecular weight fraction. GPC measurements of the molecular weight distribution of fluorescence labelled PHEA isolated from rat bile have shown a pronounced shift in the molecular weight distribution profile to lower molecular weight compared with the original sample or with the polymer accumulated in the liver (Fig. 8) [29]. In a study employing PVP, about 1.2% and 1.9% of the dose administered to rats was excreted into the bile fistula within 5 hours, when PVP samples with $\overline{M}_w = 4{,}000$ and 2,000, respectively, were used [108]. A molecular weight range between 6,000 and 10,000 has been suggested as an upper limit for biliary elimination of PVP in rats [123].

The kinetics of biliary excretion of PVP given in the study by Hespe et al. [108] does not exhibit any lag phase; the highest amount of PVP is collected within the first hour (more than 60% of the total amount excreted) and the rate gradually decreases and approaches zero after five hours. These results may suggest that the rate of biliary excretion had followed the concentration level of PVP in the blood plasma rather than the amount accumulated in the liver cells. The kinetics as well as the marked dependence on the molecular size may suggest that a substantial portion of the polymer appears in the bile mainly through diffusion through the intercellular junctions, which

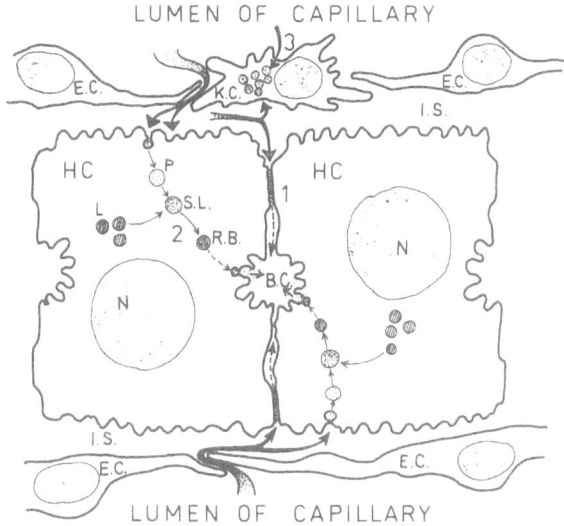

Fig. 7. Transport pathways of a polymer in the liver: 1 — diffusion through the intercellular junctions (molecular-size limited process); 2 — transcellular route of polymer transport into the bile including pinocytosis into hepatocyte and exocytosis of the vesicles or residual bodies at the lateral side of the cell; 3 — pinocytosis into the Kupffer cells occurs regularly, from either the central or the interstitial compartment. H.C. — hepatocyte; E.C. — endothelial cell of capillary wall; K.C. — Kupffer cell; I.S. — interstitial space; B.C. — bile canaliculi; P — pinocytic vesicle; L — lysosomes (primary); S.L. — secondary lysosome; R.B. — residual body; N — nucleus

are not very permeable to large molecules. It should be noted, however, that both PVP and PHEA are not accumulated in the liver mainly by the hepatocytes but by the Kupffer cells [29, 47], which, of course, cannot eliminate polymers directly into the bile (see Figs. 7, 9).

3.2.7 Mechanism of Polymer Storage in Cells

Pinocytosis seems to be the main if not the only way in which a synthetic water-soluble polymer can enter an intact cell [62]. As almost all mammalian cells have developed a pinocytic function through which they can take many important metabolites [60], they all can also capture synthetic polymers together with the surrounding fluid. The rate of polymer uptake by a particular cell is determined by the polymer concentration in the surounding medium and by the size and the rate of formation of pinocytic vesicles by the cell [60, 61].

With respect to the concentration factor, the molecules of the polymer can be found in the extracellular fluid either distributed randomly or forming some kind of concentration gradient with respect to the vicinity of the involved part of the cell surface. In the former case (fluid phase pinocytosis) [60, 61], no interactions between the polymer and the cell membrane is assumed and differences in the amount of polymer absorbed by various cells or tissues will reflect differences in their pinocytic activity. In the latter case (adsorptive pinocytosis), the higher concentration of the polymer in the vicinity of the cell surface (i.e. in a layer of a thickness comparable with the vesicle

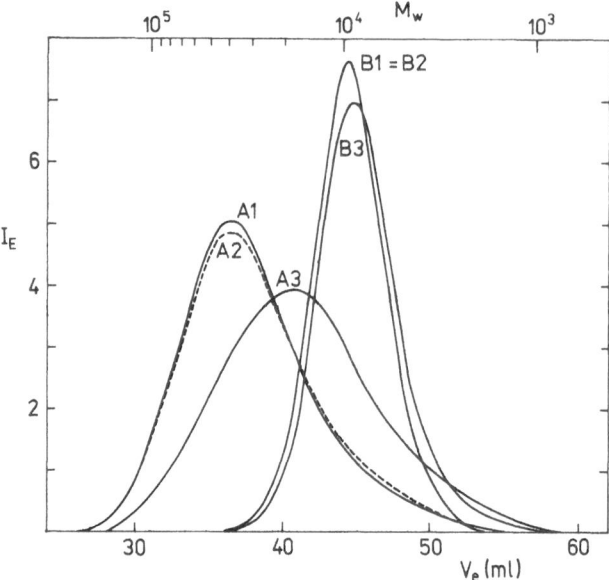

Fig. 8. GPC profiles of fluorescence-labelled PHEA reisolated from the liver tissue and the bile in rat liver perfusion experiments. A — PHEA, $\overline{M}_w = 51,400$; B — PHEA, $\overline{M}_w = 8,600$. Indices denote polymer from the perfusion liquid (1), polymer reisolated from the liver tissue (2) and from the bile (3). The curves are normalized to yield equal areas under the curve. I_E — intensity of fluorescence in arbitrary units. A pronounced shift to a lower molecular weight of the polymer from the bile indicates preferential biliary excretion of the low molecular weight fraction. The identical GPC curves of polymer from perfusion liquid and of that accumulated in the tissue indicates the uptake of polymer without any molecular size preference (Ref. [29])

size) resulting from binding of the polymer to the cell membrane may increase the rate of polymer uptake many times in comparison with the fluid phase marker [50,60,124–129]. Some, rather nonspecific interactions, e.g. electrostatic attraction of polycations to the rather negatively charged surface of the cells [125–129], or hydrophobic interactions [50,124], as well as specific ligand-receptor interactions of tailor-made polymers may be involved in adsorptive pinocytosis. This topic is discussed in more detail in another contribution to this volume [130].

On the other hand, some other factors may hinder the free diffusion of some polymer molecules to the close vicinity of the involved area of the cell surface. These factors may include steric exclusion of very large molecules due to peculiar arrangment of cell membranes [131] and electrostatic repulsion forces. Although a great deal of direct data is not available in this field, a partitioning of the polymer on the boundary between the extracellular fluid and the mucopolysaccharide layer coating the cell surface should also be considered among the factors controlling the access of the polymer to the cell membrane [119]. The glycocalyx itself, which sometimes has the character of the mucus, e.g. on cells of the intestinal epithelium, represents a layer of a rather concentrated solution of mucopolysaccharides. The phase partitioning of polymers as a function of the polymer charge, hydrophobicity and molecular weight as well as the ionic strength and the pH of the medium has often been demonstrated in vitro [132,133] for such polymers as PEG, dextran and glycoproteins. Thus, the polymer concentration in the polysaccharide layer may be much lower than in the adjacent solution [134] and, therefore, the cellular uptake of the polymer may be well below the pinocytic uptake of a fluid phase marker.

The further fate of a polymer in a cell depends on the fate of the pinocytic vesicles which may differ in particular cell types. Vesicles budding inwards from the cell membrane may travers the cytoplasm and discharge their content on another surface of the cell. This route is followed, e.g. in endothelial and epithelial cells (see above). Most incoming vesicles fuse with lysosomes, which are membrane-limited vacuoles containing a broad spectrum of hydrolytic enzymes, forming secondary lysosomes [60]. Natural macromolecules (proteins, nucleic acids, polysaccharides) are digested by the action of these enzymes to the level of low molecular weight compounds permeable through the lysosomal membrane. A polymer of limited degradability (most synthetic polymers) retains its macromolecular character and, therefore, remains enclosed in the lysosomal membrane, as this membrane is not permeable to substances with molecular weights greater than approximately 300 [135,136]. There is no evidence that synthetic polymers can escape intact from these membrane envelopes and reach other sites in the cell, although some special mechanisms probably exist for such transport of particular proteins [137,138]. Nondegradable polymers can probably be regurgitated from intact cells only by exocytosis — the reverse process to endocytosis. This mechanism has been suggested but quantitative data on the degree of its involvement are scarce. It may be inferred from the in vitro studies that in the cells, which mainly participate in the accumulation of polymers, exocytosis proceeds at a much lower rate than pinocytosis [61,139]. Therefore, polymers which are resistant to degradation in lysosomes progressively accumulate in the cells of the body and remain there for a substantial period of time. As pinocytosis seems to be a common function of almost all body cells, they all should accumulate a nondegradable polymer in this way, depending on the concentration of the latter in the surrounding medium and the pinocytic activities of the body cells. The latter differ greatly in various cells. The endocytic activities of, e.g. macrophages, endothelial cells, cells of resorptive epithelia etc., are particularly well developed and these cells play a major role in the retention of polymers in the "intracellular depo".

The reticuloendothelial system (RES) as a part of the host defense mechanism for the clearance of foreign macromolecules and particles includes several types of macrophages, which are mobile cells with particularly prominent endocytosis, able to transport endocytosed substances from the inner compartments by active motion. Kupffer cells of liver (Fig. 7, 9), reticulocytes of spleen and bone marrow, etc. are other components of RES which are in direct contact with blood. The cells of RES are responsible for a major part of the polymer accumulation from the central compartment [107,140].

It is not only the polymer circulating in the blood and lymph which can participate in the formation of the intracellular depo. The epithelial cells of kidney tubules are a very important part of the intracellular compartment (Fig. 5b). The primary filtrate formed during ultrafiltration through the capillary wall of the glomerulus is further processed in the tubulus (tube-like structures with walls consisting of a monolayer of epithelial cells) at the end of which the concentrated urine is collected. In the tubulus, mainly in the part proximal to the glomerulus, more than 99 % of the fluid filtered in

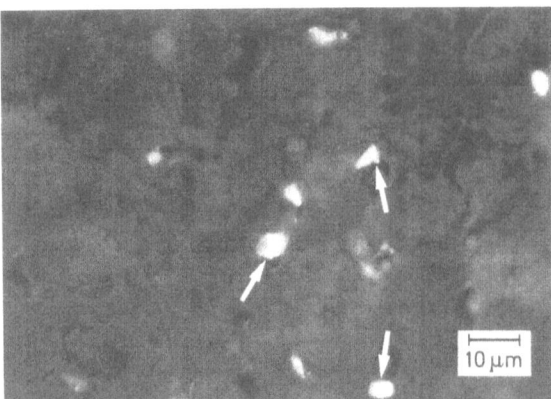

Fig. 9. Deposition of polymer I ($\overline{M}_w$ = 8,600, see Fig. 11) in the mouse liver 24 hours after intravenous administration of a dose of 3 g/kg. The only remarkable fluorescence (white spots) belongs to the polymer accumulated in the Kupffer cells (arrows) (photo F.R.)

the glomerulus is reabsorbed by the epithelial cells and transferred back into the blood. The low and middle molecular weight proteins (hormones, enzymes etc.) from the plasma of a size under the threshold of glomerular filtration, are also almost completely reabsorbed in the epithelial cells in this part of the tubulus. This reabsorption proceeds *via* pinocytosis and all the absorbed proteins undergo intralysosomal digestion [109, 141]. Because of this function, which is very important for the metabolism of low and middle molecular weight proteins, the epithelial cells of the kidney tubules belong to the family of cells with very intensive constituent pinocytosis [60]. The polymer eliminated from the plasma compartment by glomerular filtration is trapped from the filtrate in pinocytic vesicles formed during pinocytosis of proteins. If the polymer is nondegradable, it accumulates in the secondary lysosomes of the epithelial cells (Figs. 10, 12). The pronounced deposition of polymers in kidneys, which has often been observed [46,47,108,142], occurs mainly through the tubular accumulation of the polymer which is captured from the tubular lumen, i.e. from the primary urine.

Little attention has been paid to the degree of intracellular accumulation of polymers in cells of the intestinal epithelium, although a good deal of data is available indicating that accumulation, e.g. of PVP, in the intestinal wall is by no means negligible [108,113]. This question might be of great importance as the cells of the intestinal epithelium belong to the compartment with a high chemical modification capacity [143].

3.2.8 Elimination *via* the Respiratory Tract

Redistribution of the polymer in the cells participating in the formation of the "intracellular depo" may occur as a consequence of cell turnover or exocytosis of the polymer from the residual bodies (secondary lysosomes storing nondegradable material) and its recapture by other cells. Mobile macrophages of RES are particularly active in this way. Alveolar macrophages, occurring in the interior of the lung alveoli, accumulate polymers very actively [144]. The alveolar macrophages are the scavenger cells on the surfaces of the alveoli. They can pass from the lymph capillaries across the alveolar wall into the alveolar lumen and phagocytose dust and other particles from

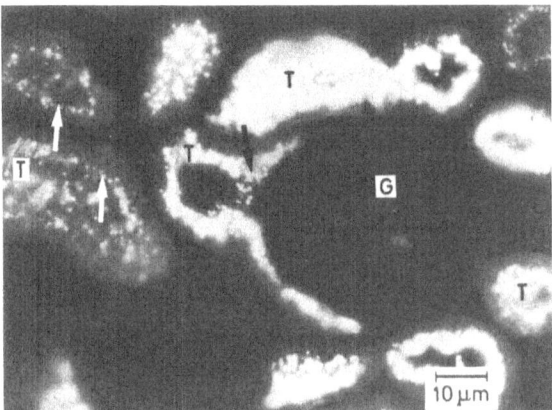

Fig. 10. Accumulation of the polymer in the cells of the kidney tubular epithelium of the mouse. The cross section through the glomerulus (G) and proximal tubules (T) in the fluorescence mode, 24 hours after intravenous administration of polymer III (see Fig. 11) at a rate of 0.1 g/kg. Fluorescence of granular deposits is seen predominantly in the apical part of epithelial cells (i.e., the part facing the tubular lumen) corresponding to the polymer accumulated in pinocytic vesicles and secondary lysosomes of these cells (arrows) (photo F.R.)

the surface of the alveoli and are then carried by cilliary movement into the larynx and swallowed or expelled in the sputum [63,145]. In this way, alveolar macrophages may also carry polymers captured during their existence in the blood or lymphatic system. This elimination via the respiratory tract may represent a way in which large nondegradable polymer molecules can leave the body. For a serious appraisal of the relative importance of the various elimination pathways already mentioned, accurate quantitative data together with molecular weight distribution analysis of the eliminated polymers are necessary. It is by no means an easy task to obtain this information.

3.3 Effect of the Nature of the Polymer and the Route of Administration on the Polymer Fate in the Body

The above discussion has revealed that, in addition to the molecular weight, the nature of the interaction of the polymer with the cell surfaces is another important factor controlling the movement of the polymer in the body compartments. Depending on the latter factor, polymers can be grouped into three classes. Polymers lacking any apparent interaction with the cell membranes, which are therefore subjected to pinocytosis with the fluid phase, belong to the first group. Polymers designed as potential plasma expanders, e.g. PVP [146], PHPMA [147], PHEA [148], dextran [149] and HES [150,151], are mostly of this type. The second group includes polymers exhibiting some kind of nonspecific interaction with the membranes of some cells, e.g. polycations [152], polyanions [153], copolymers with hydrophobic side chains [50]. The third class includes polymers containing a targeting moiety, i.e. a group, side chain or other part with some kind of molecular specificity for its counterparts (receptors) on the surfaces of particular cells [130].

Various aspects of polymer handling by a living body and many more or less general phenomena are reflected in data collected for the most studied representative of the first group, i.e. for PVP. In contrast to low molecular weight compounds, the pharmacokinetics of polymers depends more on the route of their administration. Cutaneous absorption of polymers having molecular weights of several thousands may be regarded as negligible [154]. Gastrointestinal absorption proceeds with a very low efficiency and is of little practical importance in pharmacology for polymers of molecular weights higher than approx. 40,000 [117]. Along the length of the intestine, ^{125}I-PVP was found to be best absorbed by the jejunum, with somewhat lesser amounts taken up by the duodenum and cecum; no ^{125}I-PVP was absorbed by the colon [118]. The absorbed fraction of PVP of molecular weight less than 40,000 is 0.1–0.01 % of the orally administered dose [79,117,155]. The exceedingly low gastrointestinal absorption of nondegradable polymers permits them to be used as non-toxic food additives, e.g. antioxidants, colorants etc. [117]. Taking advantage of its nondegradability, labelled PVP was tested as an indicator of increased permeability due to gastrointestinal injury occurring either under some pathological conditions or as a side effect of drug treatment [155,156]. The fraction of PVP absorbed through the gut wall enters the central compartment and its subsequent fate is similar to the fate of polymers administered parenterally [118].

The parenteral route represents the only efficient entry of a polymer into the body. This can be accomplished, e.g. by intraperitoneal, subcutaneous, intramuscular or intravascular injection of a polymer solution. However, these ways are by no means equivalent. The absorption of polymers from the cavities, such as the peritoneal or pleural, is hindered by the serose, lining the inner surface of cavities. Therefore, especially large molecules may remain there for several days [157]. The absorption is slow and takes place mainly in the lymphatic capillaries. Therefore, a rapid increase of the polymer concentration is observed in the lymphatic system, particularly in the regional lymph nodes [158], while an increase in the concentration in the central compartment and consequently rapid distribution throughout the body is protracted and delayed [85,153,159]. Owing to this behavior, which is quite different from the absorption of small molecules, a lowering in the systemic toxicity and excretion rate of basic antibiotics bound to the polyanions was observed [85].

Subcutaneous and intramuscular injections bring the polymer into the interstitial compartment and its subsequent distribution occurs through size-dependent transport via either the venous or the lymphatic capillaries.

Most of the data on the fate of soluble polymers in the living body was obtained after direct intravascular administration. Via the central compartment, a polymer is rapidly (within a few minutes) distributed to all vascularized parts of the body and its transport to other compartments starts immediately.

In the glomerular filtration, macromolecules are selectively eliminated from the central compartment into the urine in dependence on their molecular size, charge and possibly also flexibility and other properties. As synthetic polymers almost always exhibit heterogeneity in molecular size and/or composition, the distribution of the molecular parameters of the polymer remaining in the central compartment (plasma circulation) is continuously altered in time as a result of glomerular filtration. In this respect, the pharmacokinetics of the synthetic polymers is different from that of other compounds, either natural macromolecules (proteins) or low molecular weight chemicals. This fact should be borne in mind when evaluating the dynamics of polymer passage between compartments and its involvement in other physiological processes.

The kinetics of urinary excretion of PVP of various molecular weights was found to be biphasic [35]. In the first (steeper) phase, the excreted fraction represents mainly the polymer filtered directly from the plasma. The second, slower phase probably reflects the transport of the polymer from other compartments, i.e. interstitial fluid, lymph etc., to the plasma compartment. The rate of this transport is also dependent on the molecular size [75, 78]. Mainly the high molecular weight fractions can remain in the interstitial fluid for a sufficiently long time to be captured by cells *via* pinocytosis. Intracellular storage of polymers in secondary lysosomes represents the only demonstrated mode of polymer deposition in the body for a period of time ranging from months to years.

The retention of PVP, dextran, PHPMA in RES has been reported several times to be molecular-weight dependent [35, 142, 160]. Earlier studies with i.v. administered fractionated PVP showed that accumulation in typical RES organs, the liver and the spleen, is higher with fractions that have higher molecular weights [35, 160]. It was inferred that only a fraction of the large molecules present in the administered polydisperse polymer preparation is responsible for its accumulation in RES. A threshold molecular weight of 10^5 was suggested for accumulation of polymers in RES [48, 107, 160, 161]. Nevertheless, it is necessary to distinguish between the intrinsic size dependence of the pinocytic process in RES cells and size-dependent systemic effects (glomerular filtration, drainage of interstitial fluid etc.). The intrinsic effect of molecular size on the rate of polymer pinocytosis was studied in vitro with isolated peritoneal macrophages and PVP [131] and with PHEA using explanted rat liver [29]. In both cases, pinocytosis of low molecular weight fractions was sufficiently proved and no indication of preferential uptake of any molecular weight fraction was observed in the range from 0.5×10^4 up to about 10^6 (see Figs. 8, 9). Only data with PVP of a molecular weight above 7×10^6 indicated higher uptake by peritoneal macrophages which has been explained by the increased adsorptive binding to the cell surface. Therefore, it seems probable that preferential accumulation of high molecular weight samples found in vivo experiments occurs mainly as a result of indirect systemic effects which increase the relative concentration and duration of persistence of large molecules in the central compartment.

The role of molecular weight in the relative accumulation in the cells of the kidney tubular epithelium and liver RES cells can be demonstrated with data obtained for PVP. If a ^{14}C-PVP sample with a mean molecular weight of about 2,000 was administered i.v. to the rats, 92.6% and 7.0% of the administered activity was recovered from the urine and faeces, respectively, within three days [48]. Assuming the fractional uptake of PVP in the kidney tubuli to be about 0.4% of the filtered load [49], we can see that the polymer not excreted from the body should be found almost quantitatively in the kidney. The autoradiograms in the paper by Hespe et al. [48] actually show that the kidney cortex is practicaly the only place of significant polymer activity even after 24 days. When a polymer sample containing higher fraction of nonfilterable large molecules was introduced, the radioactivity of RES increased proportionaly relative to the kidney activity.

Polymers with a pronounced affinity for membranes of some cells may exhibit

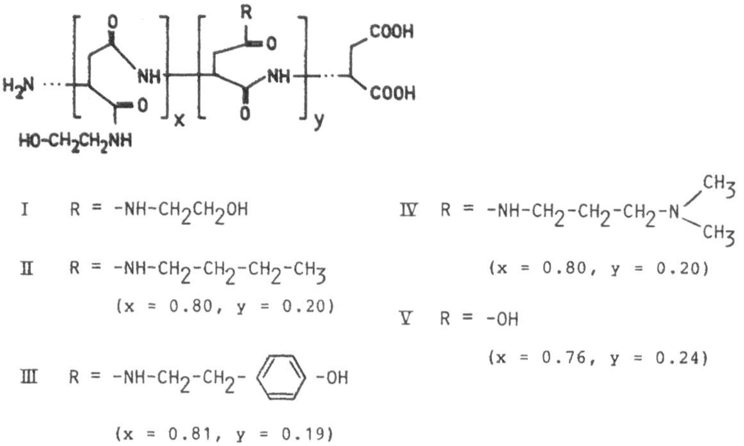

Fig. 11. Modified polyaspartamides for the study of factors modulating the adsorptive pinocytosis of polymers

a quite different distribution pattern in the body. The accumulation in the kidney tubular epithelium is of great importance as the major part of nondegradable polymer is usually eliminated *via* the kidneys, i.e. it should go through the tubuli and is thus potentially available for tubular absorption. The fractional polymer uptake (i.e. the ratio of the accumulated fraction to the total filtered polymer) depends on the polymer affinity to the membrane of the tubular cells, i.e. on the chemical structure of the polymer [50]. The fractional uptake of polymers exhibiting no apparent binding to the membranes of this type of cells is assumed to be pinocytosed proportionaly to the captured volume of the surrounding fluid, e.g. inulin [49], PVP [49], PHEA [50], and was found to be about 0.3–0.4% of the filtered load. This value may rise to more than 30% of the filtered load for copolymers containing about 20 mol-% of hydrophobic or aromatic side chains (Fig. 11) [50]. The same modification had no effect on the rate of accumulation in RES [162]. Thus, a quite different pattern of polymer distribution in tissues can be obtained after its i.v. administration (Fig. 12). It should be noted, that the role of the binding to cell membranes is emphasized in the tubular absorption of polymers, as it proceeds from the primary urine, the composition of which differs from that of the plasma or the interstitial fluid mainly in a lower concentration of proteins which can compete for the binding sites.

Polyelectrolytes have a potentiality for electrostatic interactions with the cell surfaces. The overall electric charge of the cell surface is rather electronegative and thus polycations are absorbed more strongly [125,129]. However, each type of cell exhibits its own degree of electronegativity (e.g. lymph node and spleen cells and also malignant cells are more electronegative than e.g. thymocytes or macrophages [152,163], which may lead to some level of specificity in the cellular accumulation of polycations. Some classes of polycations, e.g. quarternary polyethyleneimine, polypropyleneimine, polyvinylimidazoline, and DEAE-dextran, were tested from this point of view and they were found to inhibit tumor growth in mice [164,165]. The different affinity of spleen cells for positively or negatively charged polymers is reflected in the rate of

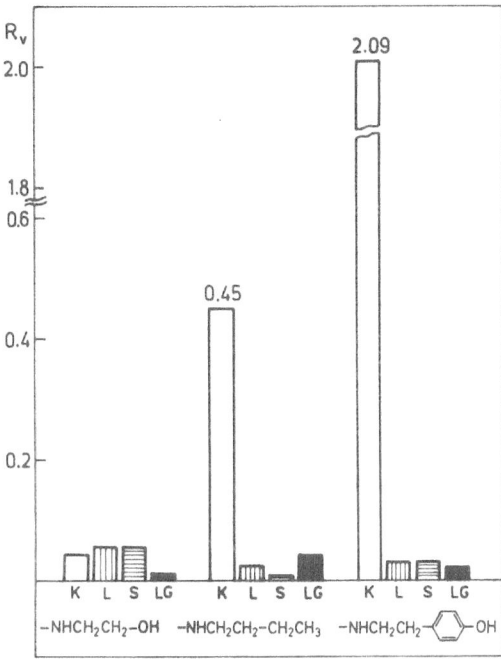

Fig. 12. The effect of various side chains on the distribution of modified poly-aspartamide polymers (I, II and III — see Fig. 11) in organs of mice 11 days after the i.v. administration of 3 g/kg in a single dose. The accumulation of polymers is expressed as the retention ratio R = (mg of polymer/g of tissue)/ (g of dose/kg of body weight), data of three animals were averaged for each value. K — kidneys, L — liver, S — spleen, LG — lung. (From Ref. [50])

disappearance (or accumulation) of polymers in the spleen of mice (Fig. 13) [162]. While the rather negative charge of most cells may explain the increased affinity for polycations, the existence of local electropositive regions could explain the adsorption of polyanions. DIVEMA, a copolymer of divinyl ether and maleic acid, was found to be captured by macrophages via adsorptive pinocytosis [166], which may be competitively inhibited by other polyanions [153].

The distribution pattern of a polymer in the body may be substantially altered by providing a polymer molecule with a substituent (ligand) exhibiting an affinity for

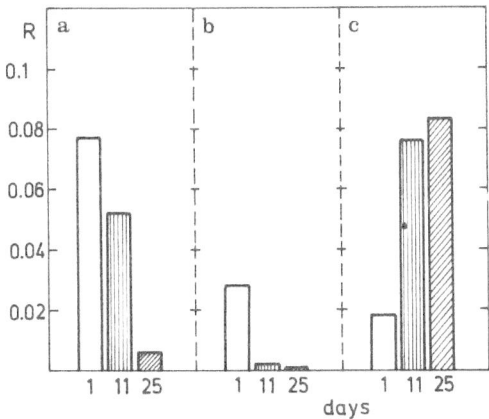

Fig. 13. Effect of the polymer charge on the affinity of the polymer for spleen cells reflected in the rate of its elimination and/or accumulation in the spleen of mice. The amounts of polymers (in R — see Fig. 12) in the spleen tissue on the day 1, 11 and 25 after i. v. administration. The amount of the polymer was determined using fluorescence labelling: a — neutral polymer; b — anionized polymer and c — cationized polymer (polymers I, V and IV in Fig.11) (from Ref. [162])

its structural counterpart (receptor) on the membranes of some cells. The idea of using such targeted polymers as drug carriers has been proposed [161,162]. It seems that there will be no problem in finding a sufficient number of ligand-receptor pairs to achieve positive adsorptive interaction with the cell membranes and thus render the cellular uptake more efficient. However, it is not easy to find ligand-receptor systems that are specific for the target tissue alone. The utilization of immunospecific determinants, e.g. antibodies to tissue-specific antigens, offers the highest known specificity but, on the other hand, the properties of the resulting conjugates are mostly determinated by the bulky protein component [162,167]. Practically, the same results could be obtained with drugs attached directly to the antibody [168]. Such "chemically modified proteins" are known to be very efficiently cleared from the plasma by the RES cells, which would be at least the second target in vivo [162,169]. The recent discovery of receptors for ligands as small as monosaccharide units [170,171], makes the idea of affinity therapy very attractive. Nevertheless, also in these cases, the modification of an "inert" carrier polymer by a targeting moiety, drug, etc. may result in a change of the overall physico-chemical properties of the macromolecule which favour the nonspecific uptake by cells exhibiting a high rate of pinocytosis (e.g. RES cells, kidney tubular epithelia, etc.). Future research in the field of absorptive interaction of polymers with cell surfaces should provide means for the compensation of undesirable sorption effects [162].

In conclusion, the macromolecular properties of polymers and their interactions with cell surfaces result in a specific pharmacokinetic behaviour of polymers. The routes of parenteral administration are far from being equivalent, e.g. the intraperitoneal application often used cannot substitute the intravenous administration. Molecular parameters of the polymer circulating in the central compartment are changed in time not necessarily by a direct biological modification of the polymer but as a consequence of a selective processing of different fractions. The intracellular accumulation in secondary lysosomes is the only proven mode of persistence of a soluble polymer in tissues. Variations in the chemical structure of the polymer may result in a different pattern of polymer distribution in the body as a consequence of a different rate of cellular accumulation.

4 Reactions in which Synthetic Polymers Can Participate

4.1 Chemical Modifications

An animal body is a chemically very active milieu, rich in a great variety of biocatalysts, reactive molecules and even radicals. As mentioned above, each of the body compartments exhibits a certain chemical activity. Liver cells are most active in this respect, followed by the kidney, RES, white blood cells, intestinal mucosa, etc. Foreign compounds which enter the body by any route except orally — called xenobiotics — are chemically modified, usually into a form that can be more easily eliminated from the body. The whole spectrum of modification reactions can by roughly grouped into the following classes:

1) Hydrolytic reactions;
2) Oxidative processes;
3) "Conjugations" — usually esterification, acylation, or alkylation.

4.1.1 Hydrolytic Reactions

Hydrolytic reactions, which are best documented, include: saponification of esters, hydrolysis of amides, and hydrolysis of ethers (glycosides). Except for ester saponification, all the other reactions proceed at a measurable rate only by enzymic catalysis. Hydrolytic enzymes are regular constituents of the digestive tract and lysosomes. They are responsible for the "biodegradation" of polyesters, polyamides (including polypeptides), polysaccharides and, probably, polyurethanes.

Natural polymers may be hydrolyzed from the ends by exoenzymes which detach the last or first monomeric unit (polarity is usually very important for the enzyme specifity). Endoenzymes, on the other hand, disassemble the polymeric chain at sites located at a certain distance from the ends. Consequently, the former mechanism leads to little changes in molecular weight; however, the low molecular weight products — represented exclusively by monomers — start to accumulate from the very beginning. The latter process causes a sharp decrease in molecular weight; however, the low molecular weight products accumulate only at late stages of degradation, being composed of a broad spectrum of oligomers. This pattern is very similar to chemical or mechanical degradation.

Disappearance, lost of weight, surface corrosion, changes in the mechanical parameters, etc. have been described with insoluble polymers implanted in the body [173]. However, there is very little direct data available on the degradation of soluble polymers in vivo. With natural polymers, the endoenzymes are usually very specific; therefore, preferential degradation from the ends can be expected with synthetic polymers, except for cases where specific sequences are inserted for purposes of degradation in the chain [130].

We used a synthetic polypeptide, α,β-poly[N-(2-hydroxyethyl)-D,L-aspartamide], which was prepared by aminolysis with aminoethanol of D,L-polysuccinimide obtained by thermal polycondensation [174]. The polysuccinimide is fully racemic and the aminolysis results in random opening of the succinimide rings, as demonstrated by the ^{13}C-NMR spectrum [175,176]. As enzymes hydrolyze only the peptide bond of L-aspartic acid, a total enzymic degradation of this polymer is not possible. If all the α-bonds of the L-isomer are hydrolyzed in the polymer, the tetrapeptide should be the final product. On the other hand, if the polymer is hydrolyzed by enzymes starting only from the chain ends (of carboxypeptidase from one and aminopeptidase from the other end), the degradation can proceed only to the D-isomer unit in the chain, and the β-bond may or may not terminate the degradation. Thus, if there are n moles of amino end groups in the polymer sample, there are $0.5n$ moles of L-units at the amino end, including $0.25n$ moles of dipeptides, $0.125n$ moles of tripeptides, etc. of all-L configuration. All these will be detached by the aminopeptidase giving a total of n moles of monomeric L-aspartic acid. If the same is true for the carboxy end, $2n$ moles can be expected provided the α- and β-bonds are hydrolysed with equal probability.

The experiment was carried out using sewage water microflora as a well known source of the broadest spectrum of hydrolytic enzymes [177,178]. Two strains of bacteria were also isolated which can grow using the polymer as the sole source of nitrogen. In both cases, no change in the molecular weight was detected by GPC, which eliminated the endoenzymic mechanism of degradation. About 4 to 6% of monomers were detached before the degradation was complete, which is equal to the fraction of the end groups. We concluded, that in our experiment, degradation proceeded from the ends of the molecule [179].

4.1.2 Oxidations

Oxidation is another very frequent process of "detoxication" of xenobiotics. It is catalyzed by an enzyme complex connected with the cytochrome P-450 and is therefore called the Cytochrome P-450 System [180]. The principal difference from the hydrolytic system consists in its particulate nature, i.e. the enzymes are bonded to intracellular cytoplasmatic structures (microsomes, cytoplasmatic reticulum, cellular membrane). Thus, it can operate only on compounds (substrates) that enter the cytoplasm.

Cytochrome P-450 is the designation of several species of haematoproteins which are conventionally defined by the affinity of their reduced form for carbon monoxide with the formation of a product with a characteristic absorption at 450 nm.

In addition to direct oxygenation, e.g. by aryl hydrocarbon hydroxylase, oxidative N- or O-dealkylation is another process catalyzed by components of the Cytochrome P-450 System (mixed-function oxidases). Reduction also occurs in this system: NADPH-cytochrome P-450 reductase has an activity similar to microsomal nitroreductase, i.e. transformation of aromatic nitro compounds into the corresponding arylamines takes place. The oxidation may be followed by other enzymic reactions, e.g. epoxides are hydrated to vicinal diols by microsomal epoxide hydratase or they are coupled with glutathione by glutathione-S-epoxide transferase.

Other types of biological oxidation are catalyzed by more specific enzymes, which play an individual role in the metabolism (various specific dehydrogenases and oxidases) or in regulatory processes (e.g. monoamine oxidase inactivating some neurotransmitters) rather than coping with xenobiotics. However, many of these enzymes are soluble and may be present in the body fluids and come into contact with components even in the extracellular compartment.

Unfortunately, we have not been able to find any published data on the effect of "detoxication" systems and of other oxidases on synthetic polymers either. This may be so because macromolecules are not subjected to these processes, or because they have not been systematically looked for. Thus, a very brief summary of these reactions is presented here in the hope that it may stimulate interest in this area.

4.1.3 "Conjugations"

The third group of xenobiotics modifications is represented by "*conjugations*", which include acylation of alcohols, phenols, amines, amides, hydrazides, etc., with sulfuric, acetic, glucuronic, glutamic and other acids, alkylation, and other synthetic reactions [180].

Glutathione-S-transfer was mentioned above as a way of modification of epoxides; however, it can also replace chlorine in chlorinated aromatic hydrocarbons. The methyl group is transferred from the S-adenosyl-L-methionine to hydrophobic thiols by thiol-S-methyl transferase, or to arylamines by arylaminemethyl transferase, or to catechols by the corresponding catechol-O-methyl transferase.

Similar to oxidations, we have not been able to find data on conjugation reactions with polymers as substrates. Nevertheless, both mechanisms of drug metabolism must be studied with polymers if the project of developing polymer-carried drugs is to be taken seriously. It is important to know whether the drug, which is coupled to the polymer, is subject to the kind of reactions known with the free drug. As most, if not all of these reactions proceed only within the cytoplasma, this study can also reveal whether a polymer can penetrate into the true interior of the cell.

4.2 Immune Reactions

Organisms are well protected against an invasion of foreign substances and bodies. They cope with small molecules by "detoxication" processes described above. The defence against particles and macromolecules is based on other principles. In addition to barriers limiting their entrance, immune reactions represent the most important mechanism. The various manifestations of the immune properties of a given substance can be classified in four groups:
1) *immunogenicity*, which is the ability to elicit the production of specific antibodies, i.e. proteins capable of binding the substance (antigen) that evoked their production;
2) *reactivity* with antibodies, which is the function of the structural determinants on the antigen (antigenic determinants);
3) induction of *tolerance* (also called immune paralysis), i.e. a state of unresponsiveness to stimulation by a particular antigen;
4) the induction of *delayed hypersensitivity* is the ability to induce specific reactivity of certain type of white blood cells against the antigen (cellular immunity) [181].

Evidence has been found that an immune reaction may be produced not only by biological macromolecules, but that the basic properties necessary to induce antibody formation are quite general to macromolecules; hence, all macromolecules can potentially elicit antibody response if given in the proper dosage and according to the proper schedule [181]. It is therefore obvious that medical or pharmacological applications of polymers may lead to serious immunological problems.

Let us consider the possibilities of synthetic polymers acting as antigens. The ability of an organism to produce an immune response depends upon the interplay between the chemistry of an antigen and the physiological state of the host. The biological mechanism leading to the immune response is very complex and various humoral factors as well as cellular interactions may take part in a particular case. However, in any case, the antigen should stimulate the antigen-sensitive cells, i.e. cells of lymphatic origin having receptors complementary to the structure of the antigen or its part (antigenic determinant) on their surface. As a result of stimulation, the cells generate an expanded population of antigen-sensitive cells and some of them initiate the production of immunoglobulins (antibodies) mirroring the specifity of the original surface receptors.

The multiple interactions of an antigenic determinant with the receptors can lead to a selection of a clone of cells (progeny of one cell) with a higher and higher affinity of receptors for the antigen. Consequently, antibodies with progressively higher specificity tend to be produced [182].

The activity of cells producing soluble antibodies is modulated by many control mechanisms which may either enhance or suppress the antibody production. While the antibody producing cells belong to the B-lymphocytes (derived from bone marrow) the control mechanisms are mostly executed by thymus-dependent T-lymphocytes, and the "helper cells" or the "supressor cells" belong to this type. The antigenic stimulation of T-lymphocytes may result not only in the control of the antibody production but also in the activation of cellular immunity, i.e. enhanced immunospecific phagocytosis and/or destruction of particulate antigen (e.g. invading or cancerous cells). The effect of polymers on this type of immune reaction, which is manifested by the delayed hypersensitivity will be discussed later.

The major thesis proposes that there is a quantitative balance between the stimulation of an immune response and the induction of tolerance following the introduction of the antigen. This balance varies for each antigen and the chemical properties of the antigen set the level and the range of dosage that can stimulate an antibody response [182].

In general, the presence of characteristic structural moieties in the molecule will enhance the immunogenicity of a polymer. The presence of aromatic side chains has the same effect, but there must be a particular density along the chain for the optimal enhancement [181]. Homopolymers, as a rule, are less immunogenic than copolymers and, among copolymers, those containing aromatic side chains are stronger immunogens [181,183,184]. The effect of the charge on the immunogenicity was examined for a series of glutamic acid-lysine (glu-lys) and glutamic acid-lysine-tyrosine (glu-lys-tyr) copolymers in which the amount of glu and lys were systematically varied. Using the same schedule of immunization and testing, the best immunogens fell in the range of $+75\%$ to -75% net charge density and within this range there was no effect of the charge on the amount of antibody formed [184]. It was inferred from studies on synthetic polypeptides that the shape of the molecule does not, in principal, affect its ability to elicit an antibody response, although it can alter either the specificity or the amount of antibodies formed. Linear polypeptides without any organized conformational structure can be potent immunogens [184], as well as synthetic polypeptides with intramolecular cross-links which have an ordered spatial structure [185]. The multichain polymers forming compact globular molecules are highly immunogenic [186]. The validity of the assumption that biodegradability is a necessary requirement for polymers to be immunogenic was tested on poly-(D-amino acids), which were found to be unsuitable immunogens. The metabolic studies suggested that the apparent lack of immunogenicity of D-amino acid polymers is due to their prolonged retention in the organs and gradual release over a long period of time, which induced immunological paralysis [184]. The use of substantially smaller doses of these polymers has led to an immune response [187,188]. Therefore, the role of polymer metabolism seems to consist in controlling the antigen level in the body. The hypothesis derived from the studies on polypeptides including poly (D-amino acids) is supported by studies on the immunogenicity of vinyl polymers.

Earlier studies with synthetic nonbiological polymers, e.g. vinyl polymers, failed to demonstrate antibody formation in animals [189,191]. This apparent lack of immunogenicity was probably due to the fact that too much antigen was administered, adopting the immunization schedule usual for metabolizable antigens [192]. An immune reaction to PVP was demonstrated in four strains of mice and each strain exhibited a somewhat different dose range for maximal response. Samples of PVP with molecular weights from 24,000 to 360,000 were immunogenic and immunological paralysis could be induced by 100 μg of polymer administered intravenously to adult mice [193]. The immunogenicity of four types of vinyl polymers was studied in rabbits, i.e. PVP ($\overline{M}_w = 180,000$), a neutral polymer; polyvinylamine (PVAm, $\overline{M}_w = 43,000$, a cationic polymer, pK = 9.4; poly(methacrylic acid) (PMA, $\overline{M}_w = 15,000$), an anionic polymer; poly(methacrylic acid-co-2-dimethylaminoethyl methacrylate) (PMD, $\overline{M}_w = 350,000$), an ampholyte polymer. PVP and PMD, i.e. neutral and ampholyte polymers, consistently elicited a moderate amount of antibody production (100 to 200 μg/ml) and uniformly charged PMA and PMAm elicited low or variable immune response (5 to 20 μg/ml) [192]. The amounts of antibody elicited by these polymers were in the same relative proportions as the amounts of antibody elicited by their polypeptide analogues. The optimal immunization dose was in the range 1 to 10 μg of polymer per animal [192]. The role of the charge was stressed in this study, although this interpretation should be considered with caution, as the polymers had very different molecular sizes.

The molecular weight plays an important role in the relative immunogenicity of small molecules, but above a certain treshold it does not appear to be a determining factor [181]. This limit is a function both of size and composition. For example, the treshold for (glu-lys) copolymers was about 30,000 to 40,000 and for (glu-lys-tyr) terpolymers it was between 10,000 and 20,000 [184]. The molecular weight dependence of vinyl polymer immunogenicity is more complex than could be explained by different pharmacokinetics of low and high molecular weight polymer alone. The immunogenicity and the stimulation of suppressor cells were studied in vivo in mice as well as in vitro with isolated lymphocytes using PVP fractions. PVP_{10} ($\overline{M}_w = 10,000$) was apparently nonimmunogenic in vivo, whereas PVP_{40} ($\overline{M}_w = 40,000$) and PVP_{360} ($\overline{M}_w = 360,000$) induced significant antibody response. The results showed, however, that PVP_{10} may be immunogenic in the absence of T-cells both in vitro as well as in vivo (in athymic mice) [194,195]. In addition, evidence was found that PVP_{10} activated suppressor cells in vivo thus also mediating the suppression of antibody response to PVP_{360} [194,195]. It should be noted that the apparent inability of some polymers to develop one type of immune response (e.g. antibody formation) should be a result of its active involvement in another type of response (immune suppression — tolerance).

In addition to biodegradability, another difference between natural and synthetic polymers should be considered. Proteins and glycoproteins — the most active antigens — are uniform in size and identical and constant in structure and shape, i.e. their sample consists of many copies of one type of molecule which does not change in time. On the contrary, the sample of synthetic polymer consists of a continuous distribution of molecular sizes, the molecules mostly being in the conformation of a flexible coil, varying the arrangement by perpetual chain motion. Originally, it was expected that, because of their uniform structure, synthetic polymers would produce homogeneous antibodies. The styrene-maleic acid copolymer (PSM) was used for checking this expectation [106-201]. Owing to the presence of aromatic and charged groups and some kind of structural organization (the polymer was found to form a rod-like structure with an axial ratio of 1:37) this polymer acts as a quite strong immunogen. In contrast to expectation, no restriction of antibody heterogeneity was found, but anti-PSM antibodies were multispecific with low affinity. While repeated contact with protein-like antigen results in the selection of a cell clone producing an antibody with progressively higher affinity, the antibodies of this late phase of immune response are not usually obtained with nondegradable synthetic polymers, as the prolonged presence of polymers produced tolerance in animals to them [200].

Another important field of immunogenicity of synthetic polymers concerns the immune response to haptens. The term hapten was introduced originally by Landsteiner in 1921 [202]. It designates any substance, large or small, which does not elicit an immune response by itself, but reacts with an antibody obtained by immunization with complete antigen. Even such small molecules as, e.g. dinitrobenzene, arsonatobenzene, fluorescein [203], etc., can serve as complete antigenic determinants eliciting the production of an antibody specific for their particular structure, if they are coupled with a macromolecular carrier [3].

Proteins, polypeptides and synthetic polymers have been found to be active hapten carriers. Both the nature of the hapten as well as of the carrier are responsible for the magnitude of the antibody response. In these terms, the group of haptens comprising the oligosaccharide units seems to occupy a particularily favoured position [3]. Among synthetic haptens, the dinitrophenyl group has probably been studied most often as a model [3,204,205]. The effectiveness of a carrier in promoting anti-hapten antibody formation is usually in agreement with the immunogenicity of the carrier itself. Proteins or complex corpuscular antigens, such as cell or cell membranes, are therefore most effective [3]. The magnitude of the immune response to DNP-protein conjugate, measured by the amount of the circulating anti-DNP antibody, may be several orders of magnitude higher than that induced by DNP side groups of a synthetic polymer (Fig. 14) [206].

On the other hand, the binding of the hapten groups to uniform synthetic polymer may create a recognizable structure on the formerly "cloudy" molecule. This may result in enhanced immune response to the carrier moiety [3,196)200].

The relatively simple and specific primary immune response to haptens on linear hydrophilic polymers, consisting of only a few types of well defined chemical subunits, makes them useful in immunological investigations. In studies on DNP-polyacrylamide systems, a general theory of the initial phase of immune response was developed [207-209]. This theory is based on the quantized model of cellular stimulation, assuming that the signal to the cell is generated by a critical number of hapten-recognizing receptor molecules clustered on the cell surface by one immunogenic molecule. Theoretical predictions derived from this model include the effect of the antigen velency and immunization dose on the balance between the magnitude of the immune response and immune suppression. Although the main implications of these studies are most important for theoretical immunology, they also help to understand a very important aspect of the behaviour of synthetic polymers in the living body, i.e. their immune reaction. Considering the polymers as drug carriers, we should also bear in mind that a drug or targeting moiety that aquires a macromolecular character in this way, may also aquire immunogen property and function like a hapten.

The specific immunological behaviour of synthetic polymers can be used in medicine in two ways: to reduce or to increase the immunological response of the organism.

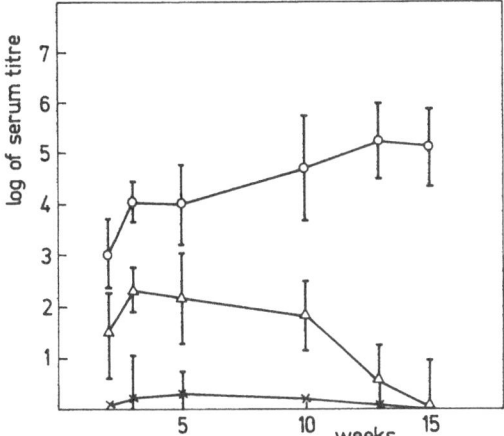

Fig. 14. Comparison of the immune response in rabbits against DNP hapten carried either by protein
(○ — bovine serum albumin (BSA), containing 18.5 mol of DNP groups/mol of BSA) or synthetic
flexible polymer (△ — PHEA with 17.3 mol- % of side chains formed by N-hexamethylene-2,4-dinitro-
aniline, X-PHEA with 2.9 mol- % of this side chain). The immune response is expressed as the titer
of the anti-DNP-antibodies circulating in the serum of rabbits after the second immunisation dose.
(From Ref. [206] together with unpublished results of autors)

Many disorders and even diseases are caused by undesirable immunological reactions:
allergy to various substances in the environment, sensitivity to some drugs, e.g. antibiotics, or even to DNA of own body (lupus erythematosus) and many others are
caused by the formation of antibodies or stimulation of cellular immunity against
these substances, presumably as a response to their combination with body proteins.
As mentioned above, combination of drugs and targeting structures with the polymer
might evoke similar undesirable side effects. In these cases, the suppression of the
immunity reaction by synthetic polymers can be utilized.

Some types of malignant cells depend on the supply of asparagine from the body
fluids. Parenteral administration of asparaginase reduces the internal asparagine pool
in the patient and curbs the growth of malignant cells. However, asparaginase is
an antigen and its administration results in the above mentioned immunological
problems. Attachment of the monomethoxy(polyethylene glycol) to the enzyme
removes its immunogenicity [210] and strongly reduces its ability to bind antibodies
if they are already present. In addition, it increases the elimination half-life from 3.35
to 32.9 hours [211]. In this particular case of asparaginase, the role of the polymer probably consists of sterical shielding of the antigenic determinants as shown by the
inability to combine with antibodies.

In other cases, advantage has been taken of the activation of suppressor cells by
the polymer. It was shown by using model haptens [212], that the suppressor cells activated by the hapten bound to a synthetic polymer (e.g., PVP or PVA) [213] are specific
for the hapten. This effect was applied to ameliorate allergy to timothy pollen [214, 215].
The allergen was coupled with PVP and the complex was used for desensibilization.
A similar approach was used with PVP to eliminate allergy to penicillins [216]. Mono-
methoxy(polyethylene glycol) has been used in conjugation with ragweed pollen antigen [217].

The synthetic polypeptide poly(D-glu-D-lys) (PDGL) [218] is particularly useful in this respect. Being composed of D-isomers, it is resistant to enzymic degradation and is very efficient in eliciting immunological tolerance even to the haptens attached to its molecule. When the benzylpenicillyl group is bound to it, the production of antibodies on subsequent sensibilization to penicillin is decreased and, more than that, the cellular immunity to penicillin is also substantially reduced [219]. The PDGL is also effective in reducing both humoral and cellular immunity to double-stranded DNA [220].

Cellular immunity, while undesirable in the cases discussed above is, on the other hand, a very powerful tool in fighting malignant cells. The activation of lymphocytes or macrophages to kill the cancer cell is a highly appreciated activity of some synthetic polymers. In addition to the induction of interferon, this is the prevailing mechanism in the anti-tumor activity of polymers [221]. Specific activation of lymphocytes, known only with natural polymers called mitogens, was also described with a synthetic polypeptide composed of (tyr-glu-ala-gly) sequences [212]. A comb-like polypeptide poly/Lys-(leu-D,L-ala)/, was suggested for the stimulation of the immunological reactivity of patients treated by antineoplastic drugs [223]. As mentioned in the next chapter, the copolymer of maleic anhydride and divinyl ether (DIVEMA) has aquired a prominent position in this connection [224, 225]. The prevention of metastases [226, 227] is particularly welcome.

4.3 Special Biological Activities

A great variety of specific biological reactions to synthetic soluble polymers has been described. As they have been summarized in excellent reviews [228–231] an attempt can be made here to analyze the mechanism of typical activities.

The most general effects are attributed to the macromolecular nature of polymers. These effects can be demonstrated on plasma expanders, which are used as an infusion to restore circulation conditions after the lost of large volumes of blood. Low osmo-molarity, viscosity, friction reduction, slow diffusion, and reduced passage across biological membranes are the most important parameters. "Biological inertness", which is required for polymers proposed as plasma expanders, indicates that any specific biological activity is unwelcome.

Naturally, a polymer free of all biological activities does not exist. PVP and some other polymers such as PHPMA, PHEA were thought to meet this demand. However, careful inspection of their deposition pattern in the body indicates accumulation in lysosomes of cells with a high rate of pinocytosis. The excessive storage of polymers may evoke an adverse reaction sometimes termed macromolecular syndrome, thesaurosis, or lysosomal storage disease [232]. It has been explained that this accumulation is a natural consequence of the basic intrinsic properties of these polymers, i.e. large molecular size and resistence to biodegradation. Even the application of only low molecular weight fractions cannot avoid their accumulation in the kidney tubular epithelia. Therefore, biodegradability becomes an imperative demand for polymers suggested as plasma expanders, and for other purposes when large and repeated doses of polymer must be used [233]. Dextran [149], HES [150, 151, 234] and derivatized gelatine [235,236)] are some of the more recently suggested polymers that could solve this problem.

On the other hand, it is possible that the biodegradability is not a necessary requirement of all applications in human medicine. A good deal of evidence is available that vinyl polymers (regarded mostly as nondegradable) may also be eliminated from cells and finally from the body. The reported results include soluble vinyl polymers [35,48,142)] as well as microparticles [237)]. In all these studies, neither degradation nor excretion of polymer in unchanged form were demonstrated. Thus, the mechanism of their elimination remains unknown. It was also reported that solely the presence of moderate amounts of the polymer in the secondary lysosomes of storage cells need not necessarily modify their physiological functions [53)]. It seems probable that biocompatibility of these polymers is a matter of quantitative balance between the rate of accumulation and the rate at which the nondegradable polymer can be cleared from the body. The possible means for this excretion were considered in Section 3.2.

Interaction with soluble components of biological system is a process which is easy to study in vitro. Chelating of divalent and multivalent cations by polyanionic polymers is a thoroughly studied reaction [238)]; however, the physiological in vivo consequences are less clear. Interactions with biopolymers are more important. Many of these interactions are based on Coulombic forces, e.g. the formation of complexes between heparin and polycations, synthetic heparinoids with proteins [239)], or ionenes complexes [240)] with negatively charged biopolymers. Another type of interaction is controlled by ligand specifity: antigenic determinants on the polymer react with specific immunoglobulins (see Sect. 4.2), or enzymes react with specific substrates of inhibitors. In vitro model studies [241,242)] revealed the effect of the polymeric backbone on the affinity to the substrate [241,243)] or inhibitor [244)] group for the enzyme. Finally, there are interactions of a still unknown nature, e.g. the effect of synthetic polymers on non-precipitating antibodies [245)].

In the evaluation of all these in vitro studies, it should be borne in mind that the biological milieu is much more complex than the model systems and usually the whole network of multicomponent interactions should be considered in vivo.

Another specifity of in vivo conditions is given by the compartmental nature of the body discussed above. Pinocytosis or phagocytosis represent a transport of the polymer in the interior of the cell only in the geometrical sense because the polymer remains physiologically isolated from the cytoplasma by the lysosomal membrane. There are techniques which can "open" the membrane for macromolecules, e.g. osmotic manipulation [246)] or fusing with liposomes [247)]. However, as far as we know, they have not yet been employed in the study of the effects of synthetic polymers in the cell interior. Although all in vitro studies of the direct interaction of synthetic polymers with nucleic acids, mitochondria, intracellular enzymes, etc., are of undeniable theoretical importance, they yield little information on the effect of these polymers in vivo.

Consequently, interaction with the cell surface is the prevailing mechanism in the biological activity of synthetic polymers. The structure of the surface of a typical animal cell was described in Section 3.2. It consists of mosaic of phospholipids, proteins and polysaccharides with specific antigenic determinants, receptors, transport proteins, etc. Negative charges are most frequent, mostly due to the sialic acid residues terminating the oligosaccharide chains. Thus, interactions with polymers are specific and non-specific. Unfortunately, these types are quite often combined, making the elucidation more complex.

The most simple non-specific interactions are due to charge effects and are, therefore, characteristic of polycations. The inhibition of bacterial growth, which is easy to measure, provides very instructive quantitative data. The early study by Katchalski [56, 248] clearly demonstrated the importance of the macromolecular form. They compared the minimum concentrations of $(L-lys)_n$ required to inhibit the growth of Escherichia coli. Values of 450, 150, 75 and 2,5 μg/ml for n = 2, n = 3, n = 4, and poly(L-lys), respectively, were found. On the other hand, a similar study with ionenes reflected the effect of charge separation. Concentrations of 128, 16, 8, 32 and 4 μg/ml were found for 3,3-, 6,6-, 2,10-, 6,10-, and 6,16-ionene, respectively [240]. Macromolecular antimicrobial drugs promise many advantages over small molecules and were discussed in the excellent review by Samor [249]. Therefore, they have been the object of an intensive research [250, 251].

The non-specific interactions are not limited to polycations. Many biological effects observed with the most studied polyanion, DIVEMA [251–255], fall into this category. This copolymer also called DVE-MA, MVE-Copolymer, or Pyran Copolymer, is a 1:2 regularly alternating copolymer of divinyl ether and maleic anhydride. It was first prepared in 1951 [251] and exhibited very interesting biological effects which have been summarized in several reviews [252–255]. The most pronounced effects are depicted schematically in Fig. 15 as a function of the molecular weight [256]. These basic patterns are also rather general for other polymers. The increase in the spleen and lung weight indicate the response to the accumulation of polymer in the RES. The rapid increase in the spleen weight corresponds to a decline in the glomerular

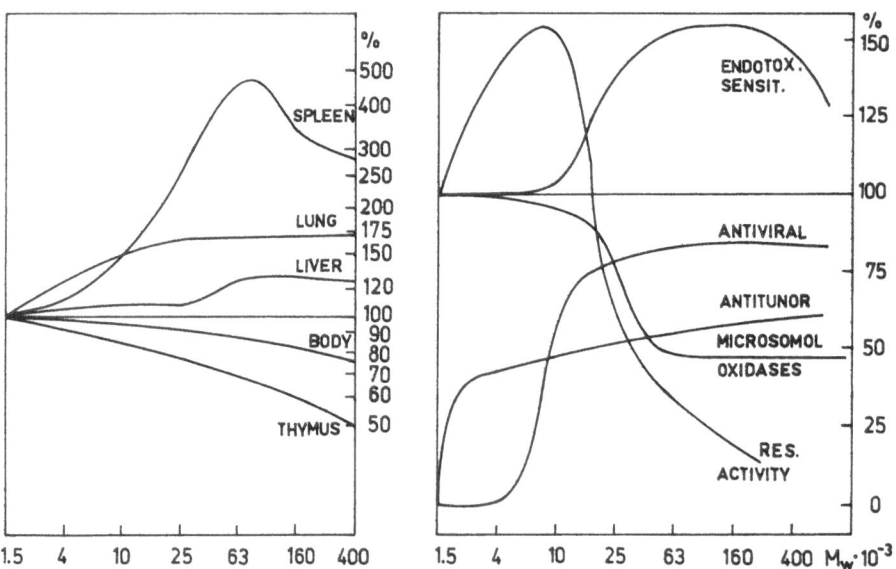

Fig. 15. Change of the general pattern of the molecular weight dependence of the organ weight and function after i.v. administration of DIVEMA 25 mg/kg in mice

filtration rate. In the same region of molecular weights, the stimulation of carbon clearance from the blood (which is a measure of the phagocytic activity of the RES) changes to inhibition, indicating the blockade of the RES function. The liver weight is affected less and an increase is observed at higher molecular weights. However, this was observed in the same region as that in which the functional changes occurred as indicated by the inhibition of mixed-function oxidases and cytochrome P-450 activity [257].

The antiviral and antitumor activity as well as the lymphocyte activation and modulation of the sensitivity to bacterial endotoxine are biological effects of DIVEMA based on the modification of the immunological responsiveness of the organism. Because of its antitumor activity, DIVEMA has been included under the code number NSC 46015 in a broad antitumor testing. Later it was found that the original NSC 46015 preparation was polydisperse and more narrow fractions were used [258]. It can be concluded from these studies that the activation of macrophages to kill the tumor cells is the predominant if not the only mechanism of antitumor activity of DIVEMA [221,259]. This is very probably true for many other synthetic polymers that have been shown to exhibit antitumor or metastase-preventing effects [260].

Another mechanism which is often claimed to be responsible for the antitumor and antiviral effects of some polymers is the induction of interferon [261,262]. Interferons [263] are glycoproteins produced by the animal cell as a response to the interaction with foreign molecules called inducers [264]. The most potent inducers are natural double-stranded RNAs which are contained in many viruses. However, many other macromolecules, mainly polyanions, are also active. Cell cultures or animal experiments can be used for testing the induction activity.

Except for synthetic polynucleotides, little data are available on the structure-activity relationship in the induction of interferon by synthetic polymers. Unfortunately, synthetic polynucleotides exhibit inherent toxicity and adverse side effects if applied in vivo.

In addition to the practical importance of interferons, there is substantial theoretical interest in the mechanism of interferon induction by synthetic polymers: it seems to be established that the inducer must penetrate inside the cytoplasma of the cell. With RNAs and polynucleotides such processes may be mediated by specific transport mechanisms which probably exist in most cells. However, it is still unclear whether other synthetic polymers can penetrate inside the cell (see the foregoing discussion).

Another type of very interesting semi-specific interaction is represented by poly-(vinylpyridine N-oxide) (PVPNO). In 1963, Schlipköter et al. [144] observed a protective effect of this polymer against silicosis. Silicosis is caused by invasion of very fine particles of silica or some other minerals into the organism. These particles are subjected to phagocytosis by macrophages resulting in the death of the macrophages followed by the loss of functional cells which are replaced by connective tissue. It was suggested that the mechanism of macrophage mortality may be found in the lysosomes where the silica particles destroy the membrane through the physico-chemical interaction, releasing lysosomal enzymes into the cytoplasma. The protective effect of PVPNO and some analogues was explained by the interaction of the polymer with the silica particles, specifically the electrons of the N-oxide with the silanol groups. If these groups are occupied by the polymer, they cannot form hydrogen bonds with the membrane and are thus prevented from destroying it [265]. It was clearly shown that the polymeric nature of the protective substance was a necessary condition [266]. There are differences between syndiotactic and isotactic PVPNO [267]. Protective activity was also demonstrated for poly(1,2-ethylenepiperidine N-oxide), poly(N,N-dimethylaminophenylene methacrylamide N-oxide) [268] and poly(dimethylamino-styrene N-oxide), whereas alcoholic and phenolic hydroxyls, amido groups, and N-acylmorpholine groups exhibited no effect [269].

However, a careful biological examination, which can be done either in vivo or in vitro by studying the direct effect of polymers on the mortality of isolated macrophages after the contact with silica, revealed that the mechanism of the protective action is probably more complex. It is known that the polymer can be administered by any parenteral route including inhalation of aerosol. Subcutaneous administration provides good protection from inhalatorily induced silicosis. In addition, an improvement is clearly apparent even if the polymer is applied four months after the development of the disease. PVPNO inhibited an increase in mesenchyme metabolism not only in silicosis but also after a systemic stimulatory effect by staphylococcal toxin [270]. Subcutaneous administration of PVPNO retarded the formation of collagen in carragenan granuloma (collagen formation indicates the proliferation of connective tissue) in guinea pigs whereas it did not influence collagen formation in normal tissue (skin slices) [271]. PVPNO is ineffective in preventing lesions caused by microparticles of chrysolite asbestos which is not toxic to macrophages [272]. Therefore, it can be concluded that the protective function of PVPNO and its analogues is not limited to interaction with the mineral particles. The physiological activity, probably the beneficial effect on the lysosomal membranes in the macrophages, plays an equally important role.

Despite the good experimental results obtained with PVPNO, polymers are not at present used in preventing and curing silicosis. Their protective effect on the macrophages is probably due to their lack of biodegradability. On the other hand, this property leads to their accumulation in the body. Table 1 lists some data obtained by Lieflander and Strecker [273] on the distribution of PVPNO in various tissues. Schmähl [274] did not find induction of malignancies even at a dose rate of 3.75 mg/kg, whereas Weller [275] observed an increase in the frequency of the malignant growth after inhalation of PVPNO. Therfore, PVPNO cannot be cleared as a safe drug for patients.

The specific interaction of polymers with the cellular surface is represented by polymer-receptor binding. Receptors are specific chemical structures with a selective

Table 1. Distribution of poly(vinylpyridine N-oxide) in organs of rats[a]

Organ	After 10 days	After 30 days
Kidney	0.11	0.18
Liver	0.23	0.16
Spleen	1.05	3.38
Lungs	0.16	0.66

[a] PVPNO was administered subcutaneously at a dose 1.5 g per animal. The data are given in % of dry tissue

affinity for one type of molecule — an analogy to the lock part in the key-lock model of enzymes. When a specific molecule is bound to the receptor, the programmed reaction is triggered. Such a "right key to the lock" is called an agonist. There may be analogues of the agonist which bind and thus block the receptor; however, they cannot initiate the proper reaction. These "wrong keys" are called antagonists. Both agonists and antagonists can be used as targeting groups in order to increase the affinity for the receptor-bearing cells. Naturally, only an agonist attached to the polymer can be expected to elicit the specific reaction.

A number of experimental problems in this field was expertly discussed by Venter [278]. Most of them fall into two categories: separation of specific and non-specific interactions and demonstration that the ligand is not detached from the polymer.

The specificity is a quantitative rather than qualitative property. There are various degrees of specificity: from a very broad to almost absolute one. The interaction of the above mentioned ionenes with the acetylcholine receptor can be given as an example of a broad specificity [265].

Acetylcholine is a neuromediator which constitutes a communication link between the nerve and muscle cells. It is discharged by the end of the neuron adhering to the muscle cell. If bound to the surface receptors, acetylcholine triggers a chain of events leading to contraction of the muscle. Thus, acetylcholine receptors are located on the surface of the cell and are accessible from the interstitial compartment. They exhibit affinity for several other substances, currare type poisons being the best known antagonists [277]. This affinity can be expressed as the concentration which competes with the binding of the testing ligand, as to reduce its binding to 50%. Data obtained with α-bungarotoxin as the testing ligand are summarized in Table 2.

Table 2. Affinity of ionenes for the acetylcholine receptor compared with a few cholinergic drugs [265]

Ligand	I_{50} [a]
Ionene 3,3	0.62
Ionene 4,6	0.57
Ionene 3,4	0.48
Ionene 3,6 and 6,8	0.40
Ionene 6,6 and 10,6	0.30
Ionene 6,16	0.24
Propionylcholine	0.7
D-Tubocurarine	0.12
Acetylcholine	3.13
Hexamethonium	2.8
Decamethonium	2.1
Nicotine	92
Atropine	65
Pilocarpine	450
Choline	1750

[a] Concentration in μ-moles. l^{-1} of the ligand reducing the α-bungarotoxin binding to 50%

The competing of ionenes with the testing ligand not only occurred at lower concentrations than with non-polymeric analogues (hexamethonium, decamethonium), but has much a steeper dependence which indicates irreversible binding.

A higher specifity is usually achieved by linking of low molecular weight ligands to polymers. Naturally, the binding site should be carefuly selected to preserve the ligand-receptor affinity. Successful attachment of an agonist to a soluble polymer with the preservation of both affinity for the receptor and physiological activity can be demonstrated by studies with catecholamines [279,280]. As the amino group is involved in the binding to the receptor, coupling of diazonium salts to the polymer with the aromatic moiety of the catecholamine was used. A similar approach was used to bind the inhibitor of trypsin and kallikrein to a similar soluble polymer [242]. In both cases, the binding to the target (to the receptor in the former and to the enzyme in the latter case) was well preserved.

The characteristic properties of the polymer carrier also affect the activity of the macromolecular derivative. A very extensive study in this field was published by Donaruma et al. [281].

The idea of polymeric drugs [167,229,282] prepared by linking a known drug molecule to a polymeric carrier is very popular. Experiments with antitumor drugs [168,283], antibiotics [284,285], and other types of biologically active molecules provide encouraging results. In addition, experiments in this field provide valuable data contributing to the understanding of the behaviour of polymers in the body.

Methotrexate (MTX) is a widely used antitumor drug which selectively inhibits dihydrofolate reductase — an enzyme involved in the synthesis of DNA building blocks. Some tumor cells, however, become resistant to it due to the blocking of MTX transport inside the cell. Suprisingly, it was found [286,287] that attachment of MTX to some polymers may overcome this resistance by an increase of the drug transport inside the cell. Many polymers were tested [288,289] and poly(L-lys) seems to be most efficient [290]. As poly(D-lys) is inactive in this respect [287,291] and as the conjugate cannot inhibit the target enzyme in vitro, it was suggested that enzymic detachment of the drug from the carrier is necessary. It was verified that this process proceeds in lysosomes [292] using inhibitors of lysosomal enzymes. Thus, polymer-bound MTX can be considered as an example of the lysosomotropic drug form [293].

The concept of lysosomotropic drug forms was suggested by the discoverer of lysosomes — Christian deDuve [62]. Studies with anthracycline antitumor antibiotics coupled to DNA [294,295] or proteins [296] demonstrated the selectivity and reduced toxicity of these drug forms. Recently, liposome entrapment has been used to render drug lysosomotropic [297]. The advantage of liposomes is in an easy decomposition and metabolism of their components.

5 The Future of Synthetic Polymers in Medicine and Pharmacology

Attempts to use synthetic polymers in medicine can be traced to the early beginnings of polymer science. Even the recent literature devoted to this topic would represent a remarkable library. Still, the utilization of purely synthetic soluble polymers in clinical practice has rarely progressed beyond the experimental phase. Nevertheless, every new achievement in polymer synthesis has led to renewed attempts to develop useful applications of synthetic macromolecules in biological and medical fields. Any success in this area depends greatly on the proper choice of a purpose which a given polymer application should fulfill.

One of the requirements very often placed on polymers designed for the use in an organism (and, at the same time, a very vaguely defined requirement) is their "biocompatibility". Biocompatibility should not be understood as an inherent property of a certain polymer but rather as a result of a dynamic process including the rate of polymer accumulation and duration of its persistence in a given compartment, its modification, and the rate of its clearance from the cells and ultimately from the body. All these variables are controlled not only by the molecular parameters of a particular polymer but also by the dose, route and frequency of its administration, which, on the other hand, are conditions dictated by the purpose of its application.

Experience with non-degradable polymers as plasma expanders indicated their accumulation in the body, which may lead to pathological side-effects. It seems now to be sufficiently well established that limited degradability is a serious hindrance in such applications of polymers, when large or repeated doses of polymers are needed, resulting in an imbalance between the uptake rate of the polymer in the cells and the capacity of elimination processes. Attempts to achieve a prolonged presence of a drug in the circulation or to obtain a depot effect by binding the drug molecule to "inert" polymers as carriers will mostly fall into the above category for drugs that are active only after detachment from the carrier. Taking into account the distribution volume, the rate of detachment and the rate of the elimination (by excretion or metabolisms) of the free drug, it can be readily calculated that the administered dose of a polymer-carried drug necessary to maintain a steady effective concentration of the free drug (e.g. antibiotic) in the circulation for several days may be tens or hundreds of grams. In addition, the effect of the polymer structure modification by the attached drug on the distribution and accumulation of the polymer-drug complex in the body appears to be a very important factor which, unfortunately, is not yet fully predictable.

Polymeric drugs active in their macromolecular form can be used at more favourable quantitative levels; thus, the accumulation-elimination balance could be more easily maintained. With this type of drugs, cell surface receptors are the most probable effector sites. Although the effects on intracellulary localized functional units such as the nucleus, nucleic acids, mitochondria, ribosomes, nuclear and microsomal enzymes, etc., have been suggested as explanations of many of the biological activities of synthetic polymers, it should be remembered that the penetration of the synthetic polymer in the cytoplasma outside lysosomes must first be demonstrated in these cases. Further research in this field could provide very valuable results, if complemented by observations of the overall biological effects with quantitative investigation of intracellular transport of the polymer and interaction with membranes. The most promising applications of polymeric drugs active in the macromolecular form could be expected in the utilization of the antimicrobial activity of certain polymers which provides a possibility of suppression of the microbial population in one compartment (e.g. the intestine) without flooding the whole body with the drug. Interferon induction, modulation of the immune reactions and control of hormone production and action seem to be other possible applications in the immediate future.

Synthetic water-soluble polymers are a particularly useful subject in the targeting of drugs. The idea of affinity chemotherapy is so attractive that it alone renders the pharmacological use of polymers sufficiently reasonable. Selective temporal accumulation in certain types of cells or specific adsorption on cell surface in the target tissue is often contemplated. However, recent clinical applications employ only DNA or liposomes as carriers. Nevertheless, it is evident that synthetic polymers can overcome the limitations inherent to the natural polymer (DNA) or to the particulate carrier (liposome), which is available only in the limited distribution volume and mainly for phagocyting cells in this region. However, progress in the pharmacology of synthetic polymers — as documented above — is not yet sufficiently complete to allow clinical applications. Targeted polymers may be even more useful in differential diagnostic, when carrying reporter groups, e.g. radioactive, x-ray-absorbing or fluorescence groups.

In conclusion, the future of synthetic soluble polymers in medicine and pharmacology probably lies in special applications of tailor-made polymers rather than in their bulk use. The empirical design of such special polymers would be an ineffective long-term process with very uncertain results. Thus, all possible information from biological disciplines should be assembled, processed and utilized in the rational design of a polymer for a given purpose. Specialists in biochemistry, cell biology, physiology, immunology, pathology, pharmacology, medical science, etc., should cooperate in this project. Besides the achievements in the polymer application in medicine and pharmacology other profits will result from this cooperation: synthetic polymers will be able to provide effective and versatile tools in biological and fundamental medical research as models of biopolymers and probes for barriers, membranes and transport mechanisms, as tools for specific manipulations of cellular processes, as hormone models, as well-defined antigens, etc. This represents another important field of utilization of synthetic water-soluble polymers in biological science. Consequently, new disciplines such as polymer biochemistry and polymer pharmacology may appear in the future.

6 References

1. Seige, K. et al.: Z. Inn. Med. 20, 669 (1965)
2. Brynda, E. et al.: J. Biomed. Mater. Res. 12, 55 (1978)
3. Leskowitz, S.: The immune response to haptens, in: Immunogenicity (ed.) Borek, F., p. 131, Amsterdam, North Holland Pub. Comp. 1972
4. Sundberg, M. W. et al.: J. Med. Chem. 17, 1304 (1974)
5. Hwang, K. J., Wase, A. W.: Biochim. Biophys. Acta 512, 54 (1978)
6. Kačena, V.: Chemical Effects of Decay of Incorporated Radioisotopes, in: Biological Effects of Transmutation and Decay of Incorporated Radioisotopes, IAEA, p. 199, Vienna, 1968
7. Oliver, R., Lajtha, L. G.: Nature 186, 91 (1960)
8. Drobník, J.: Makromol. Chem. 180, 1597 (1979)
9. Mück, K.-F., Christ, O., Kellner, H.-M.: Makromol. Chem. 178, 2785 (1977)
10. Reinhold, H. S., in: Frotiers of Radiation Therapy and Oncology (eds.) Vaeth, J. M., Karger, S., p. 44, Basel, 1972
11. Kamogawa, H.: J. Polym. Sci. A-1, 7, 2458 (1969)
12. Kamogawa, H.: Prog. Polym. Sci. Japan 7, 39 (1974)
13. Krejčoves, J., Drobník, J., Kálal, J.: Coll. Czech. Chem. Commun. 44, 2211 (1979)
14. Nishijima, Y. et al.: J. Polym. Sci. A-2, 5, 23 (1967)
15. Coons, A. H. et al.: J. Immunol. 45, 159 (1942)
16. Riggs, J. L. et al.: Amer. J. Pathol. 34, 1081 (1958)

17. Murtha, T. P.: Syntheses of Antibody Fluorochromes Derived from Coumarin, PhD Thesis, Iowa State University of Science and Technology, Ames, Iowa, 1968
18. Nairn, R. C.: Fluorescent protein tracing, London, E. and S. Livingstone Ltd, 1964[2]
19. Öbrink, B., Laurent, T. C., Rigler, R.: J. Chromatogr. *31*, 48 (1967)
20. De Belder, A. N., Granath, K.: Carbohydr. Res. *30*, 375 (1973)
21. Biddle, D.: Ark. Kemi *29*, 543 (1968)
22. Kitamura, S., Yunokawa, H., Kuge, T.: Polym. J. *14*, 85 (1982)
23. Tanaka, H. et al.: J. Phys. Chem. *71*, 2416 (1967)
24. Teramoto, A., Morimoto, M., Nishijima, Y.: J. Polym. Sci. A-1, *5*, 1021 (1967)
25. Rypáček, F. et al.: J. Polym. Sci. Polym. Symp. *66*, 53 (1979)
26. Rypáček, F., Drobník, J., Kálal, J.: Anal. Biochem. *104*, 141 (1980)
27. Wilchek, M., Spiegel, S., Spiegel, Y.: Biochim. Biophys. Res. Commun. *92*, 1215 (1980)
28. Ishibashi, N. et al.: Analytical Chemistry *51*, 2096 (1 79)
29. Drobník, J., Rypáček, F.: Preprints of 26th IUPAC International Symposium on Macromolecules, Mainz, 1979, Vol. III, p. 1318
30. Ohkuma, S., Poole, B.: Proc. Natl. Acad. Sci. USA *75*, 3327 (1978)
31. Yokomura, E., Ishikawa, Y.: Acta Haem. Jap. *36*, 821 (1973)
32. Visser, J. et al.: Quantitative immunofluorescence in flow cytometry, in: Immunofluorescence and Related Staining Techniques (eds.) Knapp, W. et al., p. 147, Amsterdam, Elsevier/ North-Holland, 1978
33. Bancher, E., Hoelzl, J., Schiffauer, R.: Protoplasma *66*, 327 (1968)
34. Dhumeaux, D. et al.: Biol. Gastro-Enterol. *1*, 34 (1968)
35. Ammon, R., Depner, E.: Z. Ges. Exp. Med. *128*, 607 (1957)
36. Hecht, G.: Arch. Exp. Path. Pharmakol. *226*, 46 (1955)
37. Hydén, S.: K. Lantbrukshegsk. Ann. *22*, 139 (1955)
38. Wilkinson, A. W., Storey, I. D. E.: Lancet *1*, 1269 (1954)
39. Levy, G. W., Fergus, D.: Anal. Chem. *25*, 1408 (1953)
40. Scott, T. A., Melvin, E. H.: Anal. Chem. *25*, 1656 (1953)
41. Fuehr, J., Kaczmarczyk, J., Kruettgen, C. D.: Klin. Wochschr. *33*, 729 (1955)
42. Bargmann, W.: Deut. Med. Wochenschr. *71*, 184 (1946)
43. Heinlein, H., Hübner, G.: Beitr. Path. Anat. *119*, 301 (1958)
44. Vilde, L. et al.: Arch. Anat. Path. *16*, A24 (1968)
45. Ammon, R., Mohn, G.: Acta Histochemica *6*, 66 (1958)
46. Mowry, R. W., Longley, J. B., Millican, R. C.: J. Lab. Clin. Med. *39*, 211 (1952)
47. Marek, H., Matzkowski, H., Koch, H.: Z. Inn. Med. *23*, 233 (1968)
48. Hespe, W., Blankwater, Y. J., Wieriks, J.: Arzneim.-Forsch. (Drug. Res.) *25*, 1561 (1975)
49. Schiller, A., Reb, G., Taugner, R.: Arzneim. Forsch. (Drug Res.) *28*, 2064 (1978)
50. Rypáček, F. et al.: Pflügers Arch. *392*, 211 (1982)
51. Neumann, D.: Acta Histochemica *57*/Supp. 17, 325 (1967)
52. Ainsworth, S. K.: J. Histochem. Cytochem. *25*, 1254 (1977)
53. Christensen, E. I., Maunsbach, A. B.: Kidney Int. *16*, 301 (1979)
54. Rypáček, F., Drobník, J.: in preparation
55. Watanabe, P. G., Ramsey J. C., Gehring, P. J.: Pharmacokinetics and metabolism of industrial chemicals, in: Progress in Drug Metabolism Vol. 5 (eds.) Bridges, J. W., Chasseaud, L. T., p. 311, New York, John Wiley and Sons Ltd. 1980
56. Katchalsky, A.: Biophys. J. *4*, 9 (1964)
57. Maggio, B., Ahkong, Q. F., Lucy, J. A.: Biochem. J. *158*, 647 (1976)
58. Pitha, J.: Nucleic acids and sulphate and phosphate polyanions, in: Anionic Polymeric Drugs (eds.) Donaruma, L. G., Ottenbrite, R. M., Vogel, O., p. 277, New York, John Wiley and Sons, 1980
59. Scholnick, P., Lang, D., Racker, E.: J. Biol. Chem. *248*, 5175 (1973)
60. Silverstein, S. C., Steinman, R. M., Cohn, Z. A.: Ann. Rev. Biochem. *46*, 669 (1977)
61. Williams, K. E. et al.: J. Cell. Biol. *64*, 113 (1975)
62. De Duve, C. et al.: Biochem. Pharmacol. *23*, 2495 (1974)
63. Janqueira, L. C., Carneiro, J., Contopoulos, A. N.: Basic histology, Los Altos, Calif., Lange Medical Publications 1977
64. Simionescu, M., Simionescu, N., Palade, G. E.: J. Cell. Biol. *67*, 863 (1975)

65. Palade, G. E., Simionescu, M., Simionescu, N.: Transport of solutes across the vascular endo-thelium, in: Transport of Macromolecules in Cellular Systems (ed.) Silverstein, S. C., p. 145, Berlin, Dahlem Konferenzen,1978
66. Simionescu, M., Simionescu, N., Palade, G. E.: J. Cell. Biol. *60*, 128 (1974)
67. Pappenheimer, J. R., Renkin, E. M., Borrero, C. L.: Amer. J. Physiol. *167*, 13 (1951)
68. Renkin, E. M., Garlick, D. G.: Microvasc. Res. *2*, 392 (1970)
69. Renkin, E. M.: Circul. Res. *41*, 735 (1977)
70. Shea, St. M., Bossert, W. H.: Microvasc. Res. *6*, 305 (1973)
71. Green, H. S., Casley-Smith, J. R.: J. Theor. Biol. *35*, 103 (1972)
72. Karnovsky, M. G.: J. Cell. Biol. *35*, 213 (1967)
73. Simionescu, N., Simionescu, M., Palade, G. E.: J. Cell. Biol. *53*, 365 (1972)
74. Grotte, G.: Acta Chirurg. Scand. Suppl. *211*, 1 (1956)
75. Vogel, G., Ströcker, H., Höller, M.: Pflügers Arch. Ges. Physiol. *279*, 187 (1964)
76. Gärtner, K., Vogel, G., Ulbrich, M.: Pflügers Arch. *298*, 305 (1968)
77. Youlten, L. J. F.: J. Physiol. *204*, 112 (1969)
78. Gärtner, K. et al.: Pflügers Arch. *343*, 331 (1973)
79. Hider, R. C., Lloyd, J. C., Wheeler, P.: J. Colloid Interface Sci. *65*, 1 (1978)
80. Rapoport, S. I.: Structure and function of the blood-brain barrier, in: Cerebral Metabolism and Neural Functions (eds.) Passomeau, J. V., Hawkins, R. A., Lust, W. D., p. 96, Baltimore, Williams and Wilkins 1980
81. Broadwell, R. D., Brightman, M. W.: J. Comp. Neurol. *166*, 257 (1976)
82. Mann, J. D.: J. Neurosurg. *50*, 343 (1979)
83. Renkin, E. M.: Pulm. Edema *1979*, 145
84. Taylor, A. E. et al.: Lymphology *6*, 192 (1973)
85. Málek, P. et al.: Antibiotics Annual, p. 546 (1958)
86. Szabó, G., Magyar, Z., Molnár, G.: Lymphology *6*, 69 (1973)
87. Kefalides, N. A.: Biochemical studies of the glomerular basement membrane in the normal kidney, in: Advances in Nephrology (eds.) Hamburger, J., Crosnier, J., Maxwell, M. H., p. 3, Chicago, Year Book, 1972
88. Blau, E. B., Haas, D. E.: Lab. Invest. *28*, 447 (1973)
89. Renkin, E. M., Gilmore, J. P.: Glomerular filtration, in: Handbook of Physiology Section 8 — Renal Physiology (eds.) Orloff, J., Berliner, R. W., p. 185, Washington, DC, 1973
90. Brenner, B. M., Deen, W. M., Robertson, C. R.: Ann. Rev. Physiol. *38*, 9 (1976)
91. Deen, W. M., Bohrer, M. P., Brenner, B. M.: Kidney Int. *16*, 353 (1979)
92. Chang, R. L. S. et al.: Biophys. J. *15*, 861 (1975)
93. Chang, R. L. S. et al.: Biophys. J. *15*, 887 (1975)
94. Chang, R. L. S. et al.: Kidney Int. *8*, 212 (1975)
95. Rennke, H. G., Patel, Y., Venkatachalam, M. A.: J. Clin. Invest. *63*, 713 (1979)
96. Jørgensen, K. E., Møller, J. V.: Am. J. Physiol. *236*, F103 (1979)
97. Deen, W. M., Satvat, B., Jamieson, J. M.: Am. J. Physiol. *238*, F126 (1980)
98. Arturson, G., Groth, T., Grotte, G.: Clin. Sci. *40*, 137 (1971)
99. Hulme, B., Hardwicke, J.: Proc. Royal Soc. Med. *59*, 509 (1966)
100. Hardwicke, J. et al.: Clin. Sci. *34*, 505 (1968)
101. Jørgensen, K. E., Møller, J. V., Sheikh, M. I.: Acta Physiol. Scand. *84*, 408 (1972)
102. Bohrer, M. P. et al.: J. Clin. Invest. *61*, 72 (1978)
103. Rennke, H. G., Patel, Y., Venkatachalam, M. A.: Kidney Int. *13*, 278 (1978)
104. Bohrer, M. P. et al.: J. Gen. Physiol. *74*, 583 (1979)
105. De Gennes, P. G.: J. Chem. Phys. *55*, 572 (1971)
106. Laurent, T. C. et al.: Eur. J. Biochem. *53*, 129 (1975)
107. Kopeček, J.: Soluble polymers in medicine, in: Systemic Aspects of Biocompatibility (ed.) Williams, D. F., p. 159, Boca Raton Florida, CRC Press 1981
108. Hespe, W., Meier, A. M., Blankwater, Y. J.: Arzneim.-Forsch. (Drug Res.) *27*, 1158 (1977)
109. Carone, F. A. et al.: Kidney Int. *16*, 271 (1979)
110. Rabkin, R., Kitabchi, A. E.: J. Clin. Invest. *62*, 169 (1978)
111. Petersen, J. et al.: Am. J. Physiol. *243*, F126 (1982)
112. Schlatter, E., Schurek, H. J., Zick, R.: Pflügers Arch. *393*, 227 (1982)
113. Marek, H., Koch, H., Seige, K.: Z. ges. exp. Med. *150*, 213 (1969)

114. Steele, R., Van Slyke, D., Plazin, J.: Ann. N. Y. Acad. Sci. *55*, 479 (1952)
115. Shaffer, C. B., Critchfield, F. H.: J. Am. Pharm. Ass. *36*, 152 (1959)
116. Smyth, H. F. Jr., Carpenter, C. P., Weil, C. S.: J. Amer. Pharm. Ass. *44*, 27 (1955)
117. Weinshenker, N. M.: Polymeric additives for food, in: Polymeric Drugs (eds.) Donaruma, L. G., Vogel, O., p. 17, New York, Acad. Press 1978
118. Yamazaki, K.: Nikon Univ. J. Med. *19*, 215 (1977)
119. Edwards, P. A. W.: British Med. Bull. *34*, 55 (1978)
120. Weatherford, H. L.: Z. Zellforsch. mikroskop. Anat. *15*, 343 (1932)
121. Hampton, J. C.: Acta Anat. *32*, 262 (1958)
122. de Duve, C., Wattiaux, R.: Ann. Rev. Physiol. *28*, 435 (1966)
123. Rozé, C., Feldmann, D., Vaille, C.: Ann. Pharmac. Franc. *29*, 513 (1971)
124. Duncan, R. et al.: Biochem. Biophys. Acta *717*, 248 (1982)
125. Shen, W.-C., Ryser, H. J. P.: Proc. Natl. Acad. Sci. USA *75*, 1872 (1978)
126. Ryser, H. J. P., Shen, W.-C.: Proc. Natl. Acad. Sci. USA *75*, 3867 (1978)
127. Pratten, M. K., Duncan, R., Lloyd, J. B.: Biochim. Biophys. Acta *540*, 455 (1978)
128. Duncan, R., Pratten, M. K., Lloyd, J. B.: Biochim. Biophys. Acta *587*, 463 (1979)
129. Davies, P. F. et al.: J. Cell Sci. *49*, 69 (1981)
130. Duncan, R., Kopeček, J.: This volume
131. Duncan, R. et al.: Biochem. J. *196*, 49 (1981)
132. Flanagan, S. D., Barondes, S. H.: J. Biol. Chem. *250*, 1484 (1975)
133. Walter, H., Krob, E., Books, D. E.: Biochemistry *15*, 2959 (1976)
134. Albertsson, P. A.: Partition of Cell Particles and Macromolecules, Stockholm, Almquist and Wiksell 1971[2]
135. Ehrenreich, B. A., Cohn, Z. A.: J. Exp. Med. *129*, 227 (1969)
136. Lloyd, J. B.: Biochem. J. *121*, 245 (1971)
137. Blobel, G., Lingappa, V. R.: Transfer of protein across intracellular membranes, in: Transport of Macromolecules in Cellular Systems (ed.) Silversterstein, S. C., p. 289, Berlin, Dahlem Konferenzen 1978
138. Neville, D. M. Jr., Chang, T. M.: Receptor mediated protein transport into cells. Entry mechanisms for toxins, hormones, antibodies viruses, lysosomal hydrolases, asialoglycoproteins and carrier proteins, in: Current Topics in Membranes and Transport. (eds.) Kleinzeller, A., Bronner, C., p. 65, New York, Acad. Press 1978
139. Roberts, A. V. S. et al.: Biochem. J. *160*, 621 (1967)
140. Pitha, J.: Polymeric drugs: effects of polyvinyl analogs of nucleic acids on cells, animals and their viral infections, in: Biomedical and Dental Applications of Polymers. (eds.) Gebelein, C. G., Koblitz, F. K., p. 203, New York, Plenum 1981
141. Maack, T. et al.: Kidney Int. *16*, 251 (1979)
142. Šprincl, L. et al.: J. Biomed. Mater. Res. *10*, 953 (1967)
143. Houston, J. B., Wood, S. G.: Gastrointestinal absorption of drugs and other xenobiotics, in: Progress in Drug Metabolism Vol. 4 (eds.) Bridges, J. W., Chaseaud, L. F., p. 57, Chicester, New York, John Wiley and Sons Ltd. 1980
144. Schlipköter, H. W., Dolgner, R., Brockhaus, A.: Deut. Med. Wochschr. *88*, 1895 (1963)
145. Ruhle, K. H. et al.: Prog. Resp. Res. *11*, 127 (1979)
146. Hecht, G., Weese, H.: Münch. Med. Wochschr. *90*, 11 (1943)
147. Kopeček, J., Bažilová, H.: Eur. Polym. J. *9*, 7 (1973)
148. Neri, P. et al.: J. Med. Chem. *16*, 893 (1973)
149. Arturson, G., Wallenius, G.: Scand. J. Clin. Lab. Invest. *16*, 76 (1964)
150. Wiedersheim, M.: Arch. Int. Pharmacodyn. Ther. *111*, 353 (1957)
151. Thompson, W. L. et al.: Infusionsther. Klin. Ernaehr.-Forsch. Prax. *6*, 151 (1979)
152. Moroson, H., Rotman, M.: Biomedical applications of polycations, in: Polyelectrolytes and their Applications (eds.) Rembaum, A., Selegny, E., p. 187, Dordrecht-Holland Boston USA, D. Riedel Publ. Comp. 1975
153. Papamatheakis, J. D. et al.: Cancer Treat. Rep. *62*, 1845 (1978)
154. Zaffaroni, A., Bonsen, P.: Controlled chemotherapy through macromolecules, in: Polymeric Drugs (eds.) Donaruma, L. G., Vogel, O., p. 1, New York, Acad. Press 1978
155. Silber, G. R., Mayer, R. J., Levin, M. J.: Cancer Res. *40*, 3430 (1980)
156. Gordon, R.: Lancet *1*, 325 (1959)

157. Punonen, R., Viinamäki, O.: Fertility and Sterility *38*, 491 (1982)
158. Rypáček, F. et al.: Preprints of IUPAC 17th Microsymposium on Macromolecules Prague 1977, p. C57
159. Noronha-Blob, L. et al.: J. Med. Chem. *20*, 356 (1977)
160. Ravin, H. A., Seligman, A. M., Fine, J.: New Engl. J. Med. *247*, 921 (1952)
161. Ringsdorf, H.: J. Polym. Sci., Polym. Symp. *51*, 135 (1975)
162. Kálal, J., Drobník, J., Rypáček, F.: Affinity chromatography and affinity therapy, in: Affinity Chromatography and Related Techniques (eds.) Gribnau, T. C. J., Visser, J., Nivard, R. J. F., p. 365, Amsterdam, London, New York, Elsevier Sci. Publ. Comp. 1982
163. Vassar, P. S.: Lab. Invest. *12*, 1072 (1963)
164. Moroson, H.: Cancer Res. *31*, 373 (1971)
165. Thorling, E. B., Larsen, B.: Acta Pathol. Microbiol. Scand. *75*, 237 (1969)
166. Pratten, M. K. et al.: Chem.-Biol. Interactions *35*, 319 (1981)
167. Goldberg, E. P. et al.: Anal. Chem. Symp. Ser. *9*, 375 (1982)
168. Wilchek, M.: Makromol. Chem. Suppl. *2*, 207 (1979)
169. Olsnes, S.: Nature, *290*, 84 (1981)
170. Neufeld, E. F., Ashwell, G.: Carbohydrate recognition systems for receptor-mediated pinocytosis, in: The Biochemistry of Glycoproteins and Proteoglycans (ed.) Lennarz, W. S., p. 241, New York, Plenum Press 1980
171. Goldstein, J. L., Anderson, R. G. V., Brown, M. S.: Nature *279*, 679 (1979)
172. Singer, S. J., Nicolson, G. L.: Science *175*, 720 (1972)
173. Williams, D. F.: Biodegradation in the human body, in: Fundamental Aspects of Biocompatibility (ed.) Williams, D. F., p. 129, Boca Raton, Fla., CRC. Press 1981
174. Vlasák, J. et al.: J. Polym. Sci., Polym. Symposia *66*, 59 (1979)
175. Pivcová, H. et al.: Biopolymers *20*, 1605 (1981)
176. Pivcová, H., Saudek, V., Drobník, J.: Polymer *23*, 1237 (1982)
177. Bunch, R. L., Chambers, C. W.: J. Water Pollution Control Fed. *39*, 181 (1967)
178. Ogata, K. et al.: Hakko Kogaku Zasshi *53*, 757 (1975)
179. Drobník, J. et al.: J. Polym. Sci., Polym. Symposia *66*, 65 (1979)
180. Jakoby, W. B. (ed.): The Enzymatic Basis of Detoxication Vol. I. and II., New York, Academic Press, 1980
181. Gill, T. J. III: The chemistry of antigens and its influence on immunogenicity, in: Immunogenicity (ed.) Borek, F., p. 5, Amsterdam, North Holland Comp. 1972
182. Mitchison, N. A.: Dose frequency and route of administration of antigen, in: Immunogenicity (ed.) Borek, F., p. 87, Amsterdam, North Holland Publ. Comp. 1972
183. Sela, M., Fuchs, S., Arnon, R.: Biochem. J. *85*, 223 (1962)
184. Gill, T. J. III, Kunz, H. W., Papermaster, D. S.: J. Biol. Chem. *242*, 3308 (1967)
185. Gill, T. J. III et al.: J. Biol. Chem. *243*, 287 (1968)
186. Sela, M., Fuchs, S., Feldman, M.: Science *139*, 342 (1963)
187. Stupp, Y., Sela, M.: Biochim. Biophys. Acta *140*, 349 (1967)
188. Jaton, J. C., Sela, M.: J. Biol. Chem. *243*, 5616 (1968)
189. Maurer, P. H.: Progr. Allergy *8*, 1 (1964)
190. Campbell, D. H.: Blood *12*, 589 (1957)
191. Campbell, D. H., Garvey, J. S.: Lab. Invest. *10*, 1143 (1961)
192. Gill, T. J. III, Kunz, H. W.: Proc. Natl. Acad. Sci. USA *61*, 790 (1968)
193. Andersson, B.: J. Immunol. *102*, 1309 (1969)
194. Zimecki, M., Webb, D. R.: Clinical Immunol. Immunopathol. *9*, 75 (1978)
195. Zimecki, M., Webb, D. R., Rogers, T. J.: Arch. Immunol. Ther. Exp. *28*, 179 (1980)
196. Lisowski, J., Wieczorek, Z., Staroscik, K.: Arch. Immunol. Ther. Exp. *20*, 731 (1972)
197. Wieczorek, Z. et al.: Eur. J. Immunol. *5*, 157 (1975)
198. Skibinski, G. et al.: Arch. Immunol. Ther. Exp. *27*, 603 (1979)
199. Skibinski, G. et al.: Arch. Immunol. Ther. Exp. *27*, 615 (1979)
200. Mikulska, J. et al.: Molec. Immun. *16*, 643 (1979)
201. Ugorski, M. et al.: Molec. Immun. *17*, 237 (1980)
202. Landsteiner, K.: Biochem. J. *119*, 294 (1921)
203. Lopatin, D. E., Viss, E. W.: Biochemistry *10*, 208 (1971)
204. Peacock, J. S. et al.: Cell. Immunol. *43*, 382 (1979)

205. Vidal-Gomez, J.: Scand. J. Immunol. *8*, 323 (1978)
206. Kálal, J. et al.: Brit. Polym. J. *10*, 111 (1978)
207. Dintzis, H. M:, Dintzis, R. Z., Vogelstein, B.: Proc. Natl. Acad. Sci. USA *73*, 3671 (1976)
208. Vogelstein, B., Dintzis, R. Z., Dintzis, H. M.: Proc. Natl. Acad. Sci. USA *79*, 395 (1982)
209. Dintzis, R. Z., Vogelstein, B., Dintzis, H. M.: Proc. Natl. Acad. Sci. USA *79*, 884 (1982)
210. Bendich, A. et al.: Clin. Exp. Immunol. *48*, 273 (1982)
211. Kamisaki, Y. et al.: Gann *73*, 470 (1982)
212. Lee, W. Y., Sehon, A. N., von Specht, B. U.: Eur. J. Immunol. *11*, 13 (1981)
213. Lee, W. Y., Sehon, A. H.: Immunol. Lett. *1*, 1 (1979)
214. von Specht, B. U., Smorodinsky, N.: Specific IgE anti-timothyimmunosupression by P4 allergen coupled to poly(N-vinylpyrrolidone), in: New Trends Allergy, Pap. Int. Symp. (eds.) Ring. J., Burg, G., p. 308, Berlin, Heidelberg, New York, Springer 1981
215. Smorodinsky, N. et al.: Immunol. Lett. *2*, 305 (1981)
216. Lee, W. Y., Sehon, A.: US Patent 4, 296.097, (1981); C.A. *96*, 11697R (1982)
217. King, T. P., Kochoumaian, L., Chiorazzi, N.: J. Exp. Med. *149*, 424 (1979)
218. Katz, D. H., Benacerraf, B.: Immunological Tolerance: Mechanisms and Potential Therapeutic Applications, New York, Acad. Press 1974
219. Chiorazzi, N., Eshhar, Z., Katz, D. H.: Proc. Natl. Acad. Sci. USA *73*, 2091 (1976)
220. Ishida, N., Sasaki, T.: Jap. Kokai Tokkyo Koho 82,42632 (1982), C.A. *97*, 17119w (1982)
221. Bartocci, A., Papademetricu, V., Chirigos, M. A.: J. Immunopharmacol. *2*, 149 (1980)
222. Lai, Ch. H., Maurer, P. H.: Cell. Immunol. *61*, 114 (1981)
223. Gaal, D., Hudecz, F., Szekerke, M.: Proc. Hung. Annu. Meet. Biochem. *21*, 31 (1981)
224. Munson, A. E., White, K. L., Klykken, P. C.: Prog. Cancer Res. Ther. *16*, 329 (1981)
225. Carrano, R. A. et al.: Prog. Cancer Res. Ther. *16*, 345 (1981)
226. Milas, L., Hersh, E. M., Hunter, N.: Cancer Res. *41*, 2378 (1981)
227. Loveless, S. E., Munson, A. E.: Cancer Res. *41*, 3901 (1981)
228. Rembaum, A., Sélégny, E. (eds.) Polyelectrolytes and their Applications, Dordrecht Holland, D. Riedel Publishing Co. 1975
229. Donaruma, L. G., Vogl, O. (eds.) Polymeric Drugs, New York, Academic Press 1978
230. Donaruma, L. G., Ottenbrite, R. M., Vogl, O. (eds.) Anionic Polymeric Drugs, New York, John Wiley & Sons 1980
231. Donaruma, L. G.: Progr. Polym. Sci., *4*, 1 (1975)
232. Hueper, W. C.: Arch. Pathol. *28*, 510 (1939)
233. Gruber, U. F. Bluterzatz: ::: Berlin—Heidelberg—New York, Springer 1968
234. Mishler, J. M. IV: Rev. Fr. Transfus. Immuno-Hematol. *23*, 283 (1980)
235. Havers, L., von Borgstede, I., Breuer, H.: Deut. Med. Wochschr. *87*, 732 (1962)
236. Zekorn, D.: Intravascular retention, dispersal, excretion and break-down of gelatin plasma substitutes, in: Modified Gelatins as Plasma Substitutes; Bibl. Haematol. 33, p. 131, Basel—New York, Karger 1969
237. Sjöhalm, I., Edman, P.: J. Pharm. Exp. Therapeutics *211*, 656 (1979)
238. Ottenbrite, R. M.: Structure and biological activities of some polyanionic polymers in ref. 230, p. 21
239. Jaques, L. B.: Heparin and related polyelectrolytes in ref. 228, p. 145
240. Rajaraman, R., Rounds, D. E.: Effects of ionenes on normal and transformed cells in ref. 228, p. 163
241. Drobník, J. et al.: Makromol. Chem. *177*, 2833 (1976)
242. Saudek, V., Drobník, J.: Makromol. Chem. *183*, 1473 (1982)
243. Kálal, J. et al.: Synthetic polymers in chemotherapy in ref. 229, p. 131
244. Havranová, M. et al.: Hoppe Seyler's Z. Physiol. Chem. *363*, 295 (1982)
245. Rypáček, F. et al.: Immunol. Lett. *4*, 49 (1982)
246. Miller, M. R., Castellot, J. J., Pardee, A. B.: Exp. Cell. Res. *120*, 421 (1979)
247. Margolis, L. B., Victorov, A. V., Bergelson, L. D.: Biochim. Biophys. Acta *720*, 259 (1982)
248. Katchalski, E., Bichowski-Slomnitzki, L., Volcani, B. F.: J. Biochem. *55*, 671 (1953)
249. Samour, C. M.: Polymeric drugs in the chemotherapy of microbial infections in ref. 229, p. 161
250. Donaruma, L. G.: Synthetic Biologically Active Polymers Progress in Polym. Sci. 4, London, Pergamon 1975
251. Butler, G. B.: J. Polym. Sci. *48*, 279 (1960)

252. Breslow, D. S.: Pure and Appl. Chem. *46*, 103 (1976)
253. Regelson, W.: Adv. Exp. Med. Biol. *1*, 315 (1967)
254. Butler, G. B.: Synthesis, characterization and biological activity of pyran copolymers, in ref. 230, p. 49
255. Fiel, R. J., Mark, E. H., Levine, H. I.: Physicochemical characteristic and molecular pharmacology of pyran copolymer, in ref. 230, p. 143
256. Regelson, W., Morahan, P., Kaplan, A.: The role of molecular weight in pharmacologic and biologic activity of synthetic polyanions, in ref. 228, p. 131
257. Morahan, P. S., Barnes, D. W., Munson, A. E.: Cancer Treat. Rep. *62*, 1797 (1978)
258. Stolfi, R. L., Martin, D. S.: Cancer Treat. Rep. *62*, 1791 (1978)
259. Dean, J. H., Padarathsingh, M. L., Keys, L.: Cancer Treat. Rep. *62*, 1807 (1978)
260. Ottenbrite, R. M. et al.: Biological activity of polycarboxylic acid polymers, in ref. 229, p. 263
261. Levy, H. B.: Polymers as interferon inducers, in ref. 229 p. 306
262. Breinig, M. C., Munson, A. E., Morahan, P. S.: Antiviral activity of synthetic polyanions, in ref. 230, p. 211
263. Friedman, R. M.: Interferons: a primer, New York, Academic Press 1981
264. Torrence, P. F., DeClercq, E.: Pharmac. Ther. *2*, 1, (1977)
265. Oscarson, D. W., Van Scoyoc, G. E., Ahlrichs, J. L.: J. Pharm. Sci. *70*, 657 (1981)
266. Dolger, R., Brockhaus, A., Schlipkoeter, H. W.: Beitr. Silikose-Forsch. Sonderband *6*, 213 (1963)
267. Holt, P. F., Lindsay, H., Beck, E. G.: Brit. J. Pharmacol. *38*, 192 (1970)
268. Aronova, G. V. et al.: Gig. Tr. Prof. Zabol. *2*, 15 (1982)
269. Ferruti, P.: Corsi Semin. Chim. 1968, 315
270. Junge-Huelsing, G., Wagner, H., Einbrodt, H. J.: Beitr. Silikose-Forsch. *94*, 41 (1968)
271. Chvapil, M. et al.: Prac. Lék. *19*, 206 (1967)
272. Davis, J. M. G.: Brit. J. Exp. Pathol. *53*, 652 (1972)
273. Liefländer, M., Strecker, F. J.: Hoppe-Seyler's Z. Physiol. Chem. *347*, 268 (1966)
274. Schmähl, D.: Arzneim. Forsch. *19*, 1611 (1969)
275. Weller, W.: Z. Ges. Exp. Med. *154*, 235 (1971)
276. Ehrlich, P.: Lancet *2*, 445 (1913)
277. Langley, J. N.: J. Physiol. (London) *39*, 235 (1909)
278. Venter, J. C.: Pharmacol. Rev. *34*, 153 (1982)
279. Verlander, M. S. et al.: Proc. Natl. Acad. Sci. USA *73*, 1009 (1976)
280. Hu, E. H., Venter, J. C.: Mol. Pharmacol. *14*, 237 (1978)
281. Donaruma, L. G. et al.: Potential structure-activity relationship indigenous to polymer systems, in ref. 229, p. 349
282. Goodaman, M., Verlander, M. S.: Polym. Prepr. *20*, 361 (1979)
283. Gross, L., Ringsdorf, H., Schupp, H.: Angew. Chem. *93*, 311 (1981)
284. Hofmann, V. et al.: Makromol. Chem. *180*, 837 (1979)
285. Panarin, E. F., Kopeikin, V. V., Afinogenov, G. E.: Bioorgan. Khim. *4*, 375 (1978)
286. Ryser, H. J., Shen, W.-C.: Proc. Nat. Acad. Sci. *75*, 3867 (1978)
287. Ryser, H. J. P., Shen, W.-C.: Cancer *45*, 1207 (1980)
288. Przybylski, M. et al.: Makromol. Chem., *179*, 1719 (1978)
289. Przybylski, M.: Cancer Treat Rep. *62*, 1837 (1978)
290. Fung, W. R. et al.: J. Natl. Cancer Inst. *62*, 1261 (1979)
291. Shen, W.-C., Ryser, H. J. P.: Proc. Natl. Acad. Asi. US *78*, 7589 (1981)
292. Chu, B. C. F., Howel, S. B.: Biochem. Pharmacol. *30*, 2545 (1981)
293. Trouet, A.: The concept of lysosomal chemotherapy: applications to neoplastic and parasitic diseases, in: Bundgaard, H. et al. (eds.) Drug Design and Adverse Reactions, p. 77, Copenhagen, Munksgaard 1977
294. Trouet, A., Deprez-deCampeneere, D., DeDuve, C.: Nature, New Biol. *239*, 110 (1972)
295. Deprez-deCampeneere, D.: Mem. Acad. Roy. Med. Belg. *47*, 213 (1974)
296. Trouet, A. et al.: Proc. Natl. Acad. Sci. US *79*, 626 (1980)
297. Trouet, A.: Bull. Mem. Acad. R. Med. Belg. *135*, 261 (1980)

K. Dušek (Editor)
Received May 31, 1983

Soluble Synthetic Polymers as Potential Drug Carriers

Ruth Duncan
Biochemistry Research Laboratory, Department of Biological Sciences,
University of Keele, Keele, Staffordshire, ST5 5BG, U.K.

Jindřich Kopeček
Institute of Macromolecular Chemistry, Czechoslovak Academy of Sciences,
16206 Prague 6, Czechoslovakia

Soluble synthetic polymers provide a potential targetable drug delivery system. In this article we discuss the consequences of the attachment of pharmaceuticals to macromolecular carriers with special reference to endocytosis and lysosomotropic drug delivery. The types of polymers which may be used as carriers are reviewed with particular regard to the methodology currently available in polymer chemistry for the synthesis of polymers bearing cell-specific targeting residues and incorporating effective polymer drug linkages. In order to be successful in drug delivery, the polymeric drug carrier must behave in a predictable and favourable manner in the biological environment. Studies concerned with the biological properties of synthetic polymers are also reviewed. The idea of using drug carriers to improve the therapeutic efficacy of pharmacological agents is receiving increasing attention, and the relationship between soluble synthetic polymers and other proposed carriers is discussed together with possible clinical applications.

Advances in Polymer Science 57
© Springer-Verlag Berlin Heidelberg 1984

1 Introduction

In recent years the use of synthetic polymers as polymeric drugs or drug delivery systems has received increasing attention, several Symposia being organised to discuss the current state-in-art of research into polymeric drugs [1], controlled release of bioactive materials [2,3], applications of immobilised enzymes and proteins [4] and the general biomedical applications of polymers [5,6,7]. Studies have largely been confined so far either to the development of sustained-release systems based on insoluble polymers in the form of powders, pellets or devices (reviewed in [8]) or to the development of polymeric drugs [1], i.e. soluble polymers which themselves display pharmacological activity.

The term "polymeric drug" includes both polymers that contain a drug or chemotherapeutic unit as part of the polymer backbone and polymers which include the active units as pendant groups or as a terminal group on the polymer chain. In the latter categories the designation "polymeric drug carrier" is more appropriate as the polymeric constituent serves simply as a drug carrier and is usually chosen because of its biological inactivity. Although the use of natural macromolecules as drug carriers has received considerable attention, the use of soluble synthetic polymers has not been extensively studied so far. The aim of this article is to outline the theoretical requirements of a synthetic polymeric drug carrier, review the methods available for synthesizing of specifically designed polymeric drug carriers and discuss the properties of polymers in relation to the biological processes which are inherently vital to the efficient functioning of a soluble, targetable drug delivery system.

The need for targetable drug carriers has long since been realised and in fact it was Paul Ehrlich in 1906 who coined the well-known phrase "the magic bullet" when describing drugs which might be selectively directed to their specific site of action [9]. During the last ten years research on drug carriers has been focused on the encapsulation of materials in liposomes which may be of therapeutic benefit. At the same time there has been ongoing research into naturally occurring macromolecules which might be used as drug carriers. It was Ringsdorf [10] in the mid-seventies who described a new theoretical model for the development of synthetic polymers which could serve as drug carriers, and the important polymer characteristics he outlined at that time provide a firm basis for all work undertaken in this area today. Before discussing those features of a synthetic polymer which are essential for its use as a drug carrier, let us first discuss why we need drug carriers and how they operate.

Most pharmacological agents are low molecular weight compounds which readily penetrate into all cell types and are often rapidly excreted from the body. Consequently, large and repeated doses must be given in order to maintain a therapeutic effect and, in addition, when specificity is limited, drugs invariably display a range of deleterious side effects. Attachment of drugs to macromolecular carriers alters their rate of excretion from the body and provides the possibility for sustained release over a prolonged period. Moreover, it limits the uptake of drug by cells to the process of endocytosis, thus providing the opportunity to direct the drug to the particular cell type where its activity is required.

In relation to drug delivery systems, pharmaceuticals may be classified into two groups, those which can initiate physiological response from outside the cell, i.e. by interaction with a membrane receptor, and those which interfere in some way with intracellular processes. The problems relating to the development of polymeric carriers for these two groups are in many cases unique. For example, drugs which have cell-surface receptors may be attached to a carrier by a non-hydrolyzable linkage providing the receptor-binding site on the drug remains accessible to the cell surface [11]. However, drugs destined for intracellular targets must ultimately be liberated from the carrier vehicle so that they reach their site of action. Although attachment of membrane-directed drugs to macromolecular carriers can produce many advantageous effects, including sustained activity and reduced toxicity, there are intrinsic difficulties in cell-specific targeting of such compounds. This review is therefore mainly concerned with polymeric carriers directing drugs which normally act via an intracellular site.

The conversion of low molecular weight drugs into high molecular weight species restricts their access to cells to the uptake processes described under the heading of endocytosis [12–14]. To understand the general rationale behind the development of drug carriers it is necessary to comprehend the endocytic pathway and terminology used by biologists for the description of the components of this system (Fig. 1). There are two different types of endocytic uptake of material by cells, phagocytosis and pinocytosis. Phagocytosis describes the capture of particulate material (usually >1 μm in diameter) by specialised cells of the reticuloendothelial system including macrophages of the liver, lung, spleen and monocytes. In contrast to pinocytosis, in phagocytosis particle attachment to the cell membrane is essential to trigger engulfment. Once a firm interaction with the cell surface occurs, the cell membrane begins to move outwards to surround the particle using the so-called zipper-like mechanism [15] as it travels over the particle surface. Eventually, the particle is engulfed within a phagocytic vacuole. Two of the most important factors relating to phagocytosis are:

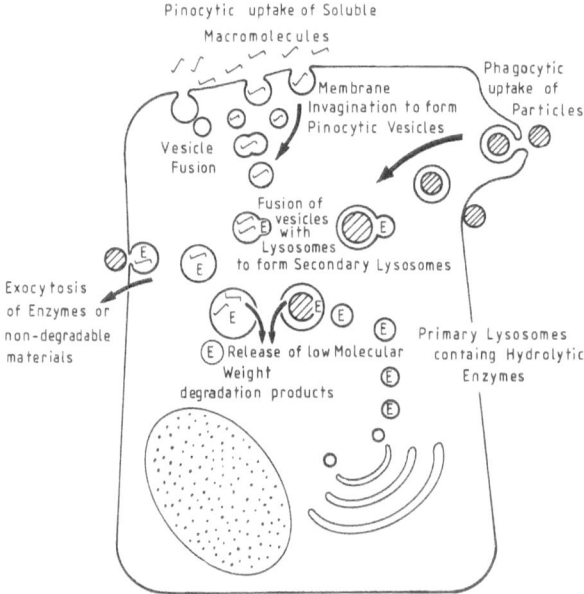

Fig. 1. Typical Endocytic Pathways. The mechanisms by which cells engulf high molecular weight particulate materials

1. the process is restricted to a few specialised cell types (phagocytes),
2. large particulate matter effectively stimulates its own uptake by such cells.

The second type of endocytosis, referred to as pinocytosis [16], describes the invagination of the cell membrane to form smaller membrane-bound vacuoles or vesicles which, during their formation, capture extracellular fluid, all solutes dissolved therein and any material adherent to the infolding surface. Pinocytosis is a phenomenon common to most, if not all, cell types and it appears to be an ongoing event with no obvious rate control mechanism. Fig. 1 shows the events which occur during and after the uptake of macromolecules by pinocytosis (the general fate of the phagocytic vacuole is the same). Subsequent to membrane invagination the newly formed vesicle pinches off and migrates into the cell cytoplasm where a series of fusion events occur. Pinocytic vesicles (pinôsomes) may fuse with each other giving rise to larger vesicles or fuse with primary or secondary lysosomes with concomitant release of membrane for recycling back to the surface [17]. Lysosomes are small vesicles of intracellular origin which contain hydrolytic enzymes [18] capable of degrading all natural macromolecules to monomeric constituents. Newly formed lysosomes are termed primary lysosomes, after fusion they are classified as secondary lysosomes. Hydrolysis of captured material proceeds within secondary lysosomes, and the low molecular weight products liberated pass through the lysosomal membrane into the cytoplasm for reutilization or removal from the cell. Non-biodegradable macromolecules accumulate within secondary lysosomes, being released slowly by exocytosis or as a consequence of cell death.

Macromolecules can be captured in two clearly different ways [19] which are illustrated in Fig. 2. When macromolecules are present in the extracellular environment, the capture of extracellular fluid inevitably causes them to become trapped and drawn into the cell. This mechanism of uptake is known as fluid-phase pinocytosis and results in the random uptake of all extracellular material. Alternatively, macromole-

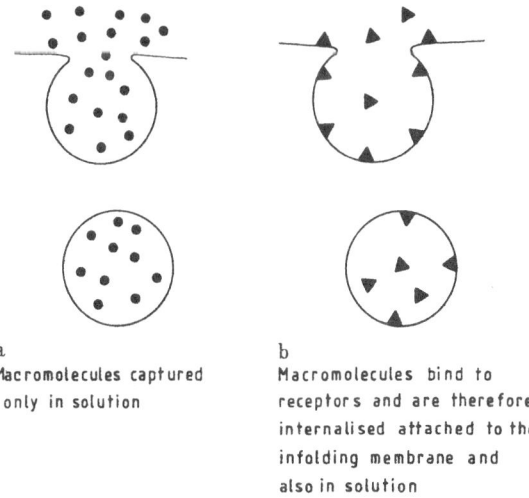

a
Macromolecules captured
only in solution

b
Macromolecules bind to
receptors and are therefore
internalised attached to the
infolding membrane and
also in solution

Fig. 2. Mechanisms of pinocytic capture of macromolecules. Uptake of substrates in solution by fluidphase (a) and adsorptive (b) pinocytosis

cules may be captured in association with the plasma membrane. This mechanism of uptake, called adsorptive pinocytosis, not only enables the concentration of macromolecules within the forming vesicles but is also responsible for the specificity of uptake. Each cell type could theoretically display a unique complement of membrane receptors and thus selectively capture specific molecules. In practice, most cells display a variety of so called non-specific receptors or binding sites which indiscriminately interact with a variety of different macromolecules. This interaction is usually caused by electric charges or hydrophobicity. However, over the past decade it has become apparent that certain cell types do indeed possess unique receptors with specific physiological functions. Most of the known pinocytic recognition systems are based on receptors on the cell surface which recognise and interact with specific carbohydrate residues present in the macromolecular substrate [20]. Systems not dependent on carbohydrate residues for recognition have also been described such as the capture of low-density lipoproteins by fibroblasts which appears to be mediated by the apoprotein B part of the low-density lipoprotein molecule and specific negative charges on the membrane receptor [21].

DeDuve and co-workers [22] first pointed out the potential value of the conversion of drugs into macromolecular species which could only penetrate cells by means of pinocytosis. They used the term "lysosomotropic drugs" to describe compounds which accumulate in the lysosomal compartment of the cell and subsequently exhibit pharmacological effects. Pinocytic uptake obviously displays several features which make it particularly amenable to utilization as a route for drug delivery. These characteristics are summarised below:

a) Pinocytic capture of macromolecules appears to be common to all cell types so that it can theoretically be harnessed to deliver chemotherapeutic agents against a multitude of different diseases.
b) The use of the known cell-specific receptor systems and those which will undoubtedly be discovered in the future would permit cell-specific targeting of pharmaceutical agents.
c) The design of drug-carrier linkages which are only cleaved intralysosomally allows the development of a carrier complex which would be stable in the extracellular environment, but release active drugs inside the cell.

Several biological consequences of pinocytic processing of a drug carrier must be borne in mind in order that the delivery system becomes effective. The interior of the lysosomes has, by nature, an acidic environment with a pH usually ranging from 5 to 6, depending on the cell type and prevailing conditions [23]; therefore, pharmacological agents which are unstable under these conditions cannot be used. Moreover, it is impossible to deliver drugs which require previous activation at some distant location such as liver microsomal activation of the antitumor agent cyclophosphamide. Success also depends on the liberation by lysosomal enzymes of a drug in a form which is not modified in such a way as to either preclude its exit from the lysosome or reduce its biological activity. Lysosomal membrane permeability has received much attention over the past years and it is known to be a phenomenon highly sensitive to small structural changes in the permeant molecules. As a general rule, molecules with a molecular weight below 200–220 easily pass through the membrane by simple diffusion [24] or active transport [25]. Of course, transport is not only affected by molecular size but also by hydrophobicity which is an important parameter determining whether or not larger molecules can pass across this barrier.

Having considered at length the fate of a polymer in the vicinity of the cell and also intracellularly following its pinocytic capture, we would like to discuss briefly the destiny of the polymer within the body. The body is itself divided into different compartments and the ability of a polymer to gain access to each compartment (and hence to the cells in that area) depends on many factors including its ability to penetrate intracompartmental barriers. Fig. 3 illustrates some of the complex interrelationships between the different compartments and some of the processes which influence the migration of polymers from one area to another. The fate of a polymer in the body is governed by a multitude of different factors, and as many factors as possible must be considered to develop an effective drug carrier.

It is obvious that any molecular characteristic of a drug-carrier conjugate can be physiologically important and should therefore be taken into account if the development of drug delivery systems is to be successful. For this reason, soluble synthetic polymers are suitable as carriers. The use of techniques available in polymer chemistry allows tailor-made polymeric carriers to be synthesized and slight modifications of the polymer composition to be performed. These modifications may yield a number of carrier macromolecules suitable for the delivery of drugs having the ability to combat a wide range of different diseases.

Polymers chosen as potential drug carriers must exhibit certain properties which are listed below in the order of their importance.

1) The polymer must be soluble and easy to synthesize ; it must exhibit a narrow, definite molecular weight distribution;

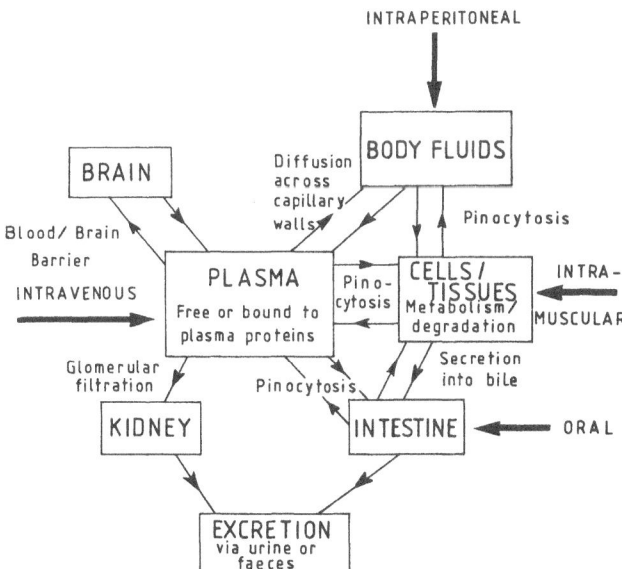

Fig. 3. Possible fate of synthetic polymers after administration to the organism

2) it should provide drug attachment/release sites or the possibility of the incorporation of drug-polymer linkages; both the sites and linkages must display controlled
   stability;
3) the polymer should display the ability to be directed *to* predetermined cell types,
   either by its inherent physico-chemical properties or the incorporation of specific
   residues;
4) it should be compatible with the biological environment, i.e. non-toxic, non-
   antigenic or not provocative in any other respect;
5) ideally, it should be biodegradable or eliminated from the organism after having
   fulfilled its function.

Unless a carrier conjugate can fulfill the prerequisities listed under (1), (2) and (3),
its usefulness is severely limited. Only when a carrier meets requirements (1)–(3) is
it necessary to consider the clinical implications of (4) and (5). Fig. 4 illustrates the
basic requirements of the polymeric carrier and the characteristic of both a linear and
crosslinked polymer. In the following section examples of soluble synthetic polymers
specifically synthesized with respect to biomedical applications are given and the
importance of their properties relative to polymeric drug carriers is discussed.

(i) Single Chain Polymeric Carrier

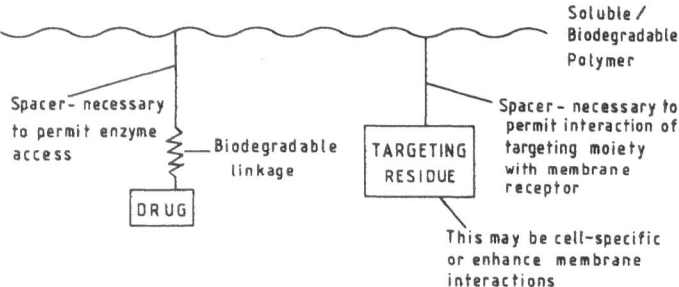

(ii) Crosslinked Polymeric Carrier

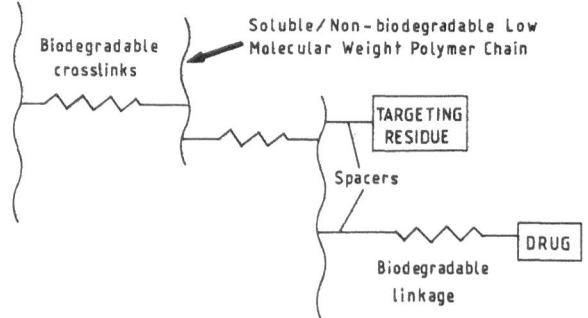

Fig. 4. Essential features of a single-chain and crosslinked polymeric drug carrier

2 Synthesis of Synthetic Polymers which are Suitable Drug Carriers

The consequences which arise from binding low-molecular weight drugs to macro-molecular carriers have already been discussed. The preparation of the ideal polymeric carrier is a complicated task which requires optimization of a number of properties of the polymer that are important from the biological standpoint. These include molecular weight, biodegradability of the backbone, possibility for drug attachment/release and for targeting, regulation of non-specific binding (charge-hydrophobicity effects), and toxicity.

The first decision that must be made in the design of polymeric drug carriers is whether to choose a synthetic or natural polymer.

These two groups of carriers mainly differ by their biodegradability and their susceptibility to structural modifications of the carrier molecule.

Natural polymers such as starch or gelatin contain many biodegradable bonds, and for many applications the structure of these polymers must be modified so as to reduce their rate of biodegradation [26]. If, on the other hand, a synthetic polymer is used which does not contain bonds degradable in the organism (e.g. poly(vinylalcohol), polyvinylpyrrolidone), the most important parameter determining its application is an appropriate molecular weight distribution.

Synthetic polymers are advantageously used as drug carriers since they are clearly more susceptible to modifications than natural macromolecules. Natural polymers usually have a limited number of functional groups suitable for binding a drug [27].

It is known that molecular weight is an important parameter in the determination of the biological activity of a polymer. For example, the rate of elimination of the polymer from the blood stream, the deposition in organs [28] as well as the rate of uptake of the polymer into the cells by pinocytosis [29, 30] are influenced by both molecular weight and molecular weight distribution. Since synthetic polymers are polydisperse, not only the average molecular weight but also the molecular weight distribution is important.

The molecular weight distribution (the polydispersity of the polymer) is usually defined by the heterogeneity index $\overline{M}_w/\overline{M}_n$ (i.e. the ratio of the weight average molecu-lar weight ($\overline{M}_w$) to the number average molecular weight ($\overline{M}_n$)) [31]. If all the molecules of a sample have the same size (natural polymers), this value is 1. The higher this value, the broader the distribution of molecular weights. The molecular weight distribution formation of an enzyme-substrate complex. Therefore, the degradable sequence of merization $\overline{M}_w/\overline{M}_n$ usually lies between 1.05 and 1.2. However, most hydrophilic polymers which are suitable as drug carriers, are prepared by free-radical polymeri-zation. The heterogeneity index of polymers prepared by this type of polymerization usually lies between 1.5 and 2.5. Of course, polymer samples with a narrower mole-cular weight distribution may be prepared by further fractionation (gradual precipi-tation). This is, however too time-consuming and usually only peformed when the highest molecular weight fractions are removed to avoid adverse side effects.

Synthetic polymers with a C—C backbone are generally not biodegrable. If they are degraded at all, degradation is so slow it may be considered as negligible [32]. How can a synthetic polymer be made enzymatically degradable? The active site of an enzyme must undergo a number of interactions with the substrate leading to the formation of an enzyme-substrate complex. Therefore, the degradable sequence of an artificial substrate must display characteristics which in some way should resemble those of the physiologically active substrate [33]. Therefore to increase the enzymatic degradability of a synthetic polymer with a carbon backbone it is usually necessary to combine it with peptide, saccharide or nucleotide sequences. This can be achieved by linking relatively short synthetic polymer chains with each other using bonds susceptible to enzymatic attack, i.e. by additional crosslinking of synthetic polymers to a level short of the gel point [34,35,36].

The advantage of a polymer thus prepared is that it contains the smallest number of bonds susceptible to enzymatic attack and that the rate of scission of these bonds can be controlled by changing the structure, amount, and length of the crosslinks. Alternatively, biodegradable synthetic polymers are obtained by synthesizing analogues of natural polymers, e.g. of synthetic poly(L-α-amino acids) [37].

A polymeric drug carrier should contain functional groups such as COOH, CHO, NH_2, and OH which allow the attachment of drugs and targeting moieties. The binding reactions should be simple and, if possible, without side reactions [38].

Copolymerization or polymeranalogous transformations [39] can be used to affect the drug distribution along the polymer chain, as well as to regulate the lipophilicity or hydrophilicity of the whole polymeric carrier molecule or of its blocks. The properties of the microenvironment in the polymer coil may differ strongly from those in solution [40,41].

By the incorporation of ionogenic commonomers and/or by a polymeranalogous reaction (e.g. hydrolysis), it is also possible to regulate the charge of the polymeric carrier. Due to the presence of ionogenic groups non-specific interactions and an change in the shape of the coil may occur (e.g. with a variation of pH and ionic strength). A change in the hydrodynamic volume may affect a number of important biological properties such as renal filtration.

The toxic effects of synthetic polymers include inhibition of hepatic microsomal oxidase enzymes, stimulation of liver transaminases, hepatosplenomegaly, thymic involution, anaemia, and decrease in bone marrow cells [42]. Fortunately, many neutral synthetic polymers do not show toxic effects. For those polymers which exhibit toxic effects, e.g. polyanions, it is sometimes possible to separate the toxicity from antitumour, antiviral and immunological effects [43]. In general the toxicity of polyanions increases with molecular weight (particularly for $M_w > 50,000$).

2.1 Types of Available Polymers

How can we control the synthesis of drug carriers to govern the above features? It has already been stated that the majority of water-soluble polymers are prepared by free-radical polymerization the mechanism of which is well-known. On the whole, the average molecular weight can be controlled relatively easily by the initiator concentration in the polymerization mixture, the polymerization temperature and conversion. In addition, the molecular weight distribution can be controlled to obtain a narrower range. If necessary, it is adjusted by fractional precipitation.

Let us review some of the polymers which have been investigated and discuss the effect of some chemical and physical parameters on their biological properties.

The effect of molecular weight distribution of polyvinylpyrrolidone on its biological properties has been described by Hespe et al. [44]. They compared two samples of this polymer having the same average molecular weight. One sample (A) contained 5% of polymer with $M_w > 25{,}000$, the other one (B) only 0.5% of macromolecules of this size range. Although both polymers were excreted at approximately the same rate after intravenous administration, polymer A was detectable in the organs during a much longer period than polymer B.

Similar results were obtained with a divinyl ether — maleic anhydride (Pyran) copolymer [45]. It was shown that the high molecular weight fractions of this copolymer are more toxic than lower molecular weight fractions.

Mück et al. [46,47] described the influence of molecular weight and tacticity of polymers on their biological properties, using a number of fractions of poly(acrylic acid)s and poly(methacrylic acid)s with different tacticity, molecular weight, and molecular weight distribution. They showed that isotactic poly(acrylic acid)s exhibit a considerably higher antiviral activity than atactic ones. Independent of the tacticity, poly-(acrylic acid) samples with molecular weights smaller than 5000 showed no significant effects. An optimum activity was observed for polymers with average molecular weights between 6,000 and 15,000. Isotactic poly(acrylic acid) samples with narrow molecular weight distributions were more active than those with broad distributions. Atactic poly(acrylic acid)s with lower efficacy do not show this relationship. Atactic poly(methacrylic acid)s do not display antiviral activity in vivo; isotactic ones produce an activity which is appreciably lower than that of atactic poly(acrylic acid).

A number of potential drug carriers are described in the literature, e.g. polyphosphazenes [48,49], derivatives of poly(methacrylic acid) [28,35,50-52], synthetic poly(amino acids) [53,54,55], polyvinyl analogs of nucleic acids [56,57], poly(ethylene oxide) [58], copolymers of vinylpyrrolidone [59], poly(amido-amines) [60], and others. The structure of these polymers has been modified by the incorporation of hydrophobic units [61,62,63], sugar residues, sulfinyl groups to achieve, in a non-specific way, affinity for certain tissues.

Hoffmann et al. [51] synthesized poly[2-(methylsulfinyl)ethyl acrylate] and poly[2-(methylsulfinyl)ethyl methacrylate] in order to examine their ability

$$R = H \text{ or } CH_3$$

to enhance the penetration of pharmaceutical agents through the skin (based on the rationale that DMSO accelerates the transdermal transport of a number of drugs). Both polymers showed a good local compatibility and the methacrylate polymer did not cause acute toxicity at systemic levels of application. However, a marked enhancement of penetration was not observed.

The same group [64] prepared [^{14}C] labelled poly[2-(methylsulfinyl)ethyl acrylate] and studied the effect of molecular weight on storage as well as the tumour affinity of the polymer. They observed a tendency for increased storage of polymers with higher molecular weight in organs containing cells which display a high endocytic activity. Specific tumour affinity (Walker carcinoma, Yoshida sarcoma) of these polymers was negligible.

Pitha [56] reviewed recently the biological properties of polyvinyl analogs of nucleic acids such as poly(9-vinyladenine) (polyVA) and poly(1-vinyluracil) (polyVU).

Poly VA   Poly VU

These polymers were originally synthesized to study interactions at the molecular level, i.e. interactions of polyvinyl analogs of nucleic acids with nucleic acids and enzymes, but their interactions with cells and tissues was also extensively investigated.

It is known that the activity of the lysosomal enzyme β-glucosidase is selectively enhanced in tumour cells by previous administration of glucose. Glucose causes a specific decrease of the pH in the tumour tissue. Carpino et al. [52] synthesized a polymeric glucopyranoside derivative of methacrylic acid in order to determine whether high molecular weight β-D-glucosides undergo accelerated degradation in tumour tissue, as has been shown with low molecular weight β-D-glucopyranosides

poly[1-0-(4-methacryloylaminophenyl)-β-D-glucopyranoside]

Duncan et al. [62,63] have shown that the incorporation of phenolic residues into macromolecules increases their rate of pinocytic capture. The incorporation of 20% tyramine residues into poly[α,β-(N-hydroxyethyl)-DL-aspartamide)] greatly increases the rate of pinocytosis of the modified polymer by rat visceral yolk sacs cultured in vitro [62].

A study of copolymers of N-(2-hydroxypropyl)methacrylamide modified by different amounts of tyrosinamide [63] has shown that a correlation exists between the percentage of tyrosinamide and the rate of pinocytic uptake of modified polymers.

Tyrosinamide units in N-(2-hydroxypropyl)methacrylamide copolymers were bound either directly to the polymer chains (A) or via a spacer (Gly-Gly) (B). The rate of pinocytosis was not affected by the method of binding. Quite understandably, the binding via a dipeptidic spacer affects the degradability of the side chain (release of tyrosine).

Some of the studies mentioned above have shown that the structure of synthetic polymeric carriers can be designed in such a way as to control their non-specific affinity for membranes. However, these interactions between macromolecular carriers and the components of body fluids or membrane surfaces are of low specificity [11]. In order to transform drugs into "magic bullets", specific interactions are needed. In other words, more sophisticated carriers with modified structures (cf. paragraph 2.3) which allow specific interactions with respective receptors have to be prepared and studied.

2.2 Polymer-Drug Linkages

The drug-polymer bond may be either covalent, when the bond strength is ~ 100 kcal/mol, or both species may be linked by weaker types of bonds, e.g. ionic bonds (5 kcal/mol), or by dipole (van der Waals) forces. Covalent bonds are regarded as the most important bonds of synthetic soluble polymers. In cases where binding occurs other than through covalent fixation, slight changes in the shape of a polymer coil may result when the polymer passes from one body compartment to another one; this can lead to dramatic changes in its biological activity. For this reason, this review is mainly concerned with covalent attachment of drugs to polymeric carriers. The drug can either be part of the polymeric backbone or bound to a side chain.

2.2.1 Polymers Containing the Drug as Part of the Polymeric Backbone

A number of chemical reactions can be employed to incorporate drugs into the polymeric backbone. These are illustrated by the following examples:

Polymerization of amino acid N-carboxyanhydrides [65], e.g. of 3,4-dihydroxyphenylalanine (DOPA), leads to the formation of polymers which possess biological activity

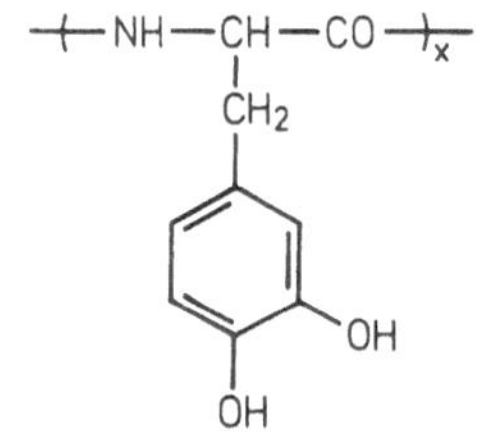

The drug can also be incorporated into the polymer backbone via a polymerizable double bond contained in the drug. In contrast, when the polymerizable group (e.g. a vinyl group) is introduced into the structure of the drug before polymerization the drug is subsequently regarded as being bound through a side chain. Pitha et al. [66] copolymerized alprenol with acrylamide. The incorporation of alprenol into a copolymer decreases its ability to bind to membrane beta-adrenergic receptors by a factor of thousand.

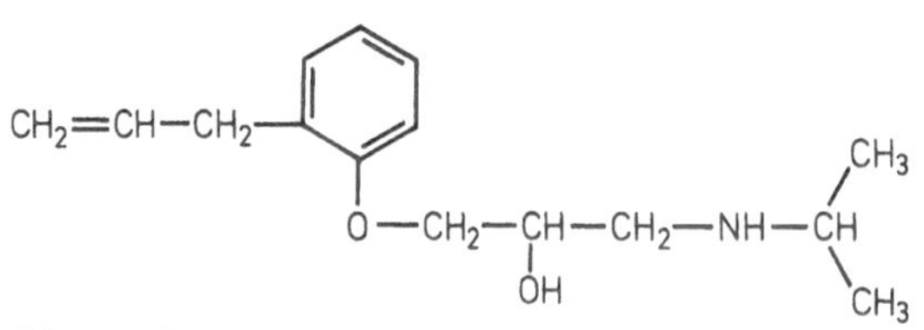

Alprenol

However, its ability to bind to antibodies specific to catecholamines and alprenol-related drugs decreases only by a factor of three.

A number of polymers which contain various metals as part of the polymeric back-bone have been prepared [67]. For example, polymer analogues of the known anti-neoplastic drug, cis-dichlorodiamineplatinum, were prepared to reduce the side effects of the low molecular weight form (gastrointestinal, imunosuppressive and renal disfunctions).

$$K_2PtCl_4 + H_2N-R-NH_2 \longrightarrow \quad \begin{matrix} Cl & Cl \\ \diagdown & \diagup \\ Pt \end{matrix} -NH_2-R-NH_2 \rightarrow_x$$

A wide variety of diamines including aromatic, aliphatic and pyrimidine diamines were used [68].

Arsenic-containing antibacterial polymers [69] have been prepared by the reaction of triphenylarsenic dichloride with 2,4-diamino-6-mercaptopyrimidine.

2.2.2 Drugs Bound to the Polymer via a Side Chain

Generally, two procedures are applied for the preparation of drugs bound to the polymer via a side chain:

a) Reaction of a low molecular weight compound containing a polymerizable group with the drug yields a derivative from which either a homopolymer drug or copolymer can be prepared.

b) Polymer analogous reaction, i.e. attachment of the drug by reacting it with the poly-meric carrier. The group with which the drug reacts may be part of the fundamental polymeric unit or introduced into the polymer by copolymerization with a comonomer containing functional groups. In addition, it is possible to introduce functional groups into a polymer chain by polymeranalogous reaction; subsequently, binding of the biologically active compound to the carrier occurs in the second polymeranalogous reaction.

These procedures can be demonstrated schematically as follows:

a) i) synthesis of

$$\begin{matrix} CH_3 \\ | \\ CH_2\!\!=\!\!C-CO-NH-CH_2-CO-NH-Drug \quad I \end{matrix}$$

ii) copolymerization

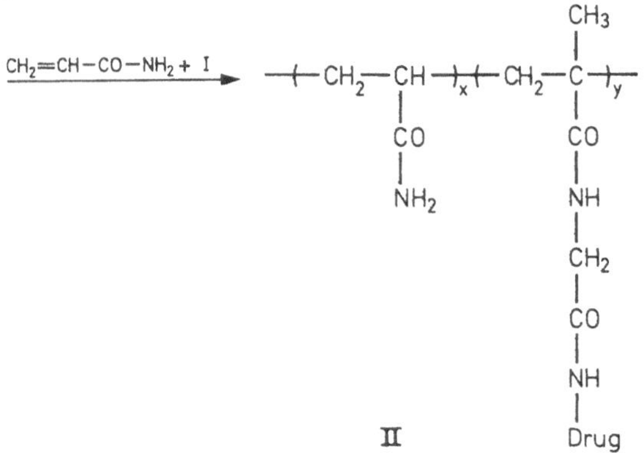

$$CH_2{=}CH{-}CO{-}NH_2 + I \longrightarrow$$

$$+CH_2{-}\underset{\underset{NH_2}{\overset{|}{\overset{CO}{|}}}}{CH}{\to}_x{+}CH_2{-}\underset{\overset{|}{CO}}{\overset{CH_3}{\overset{|}{C}}}{\to}_y$$

with the side chain: CO — NH — CH₂ — CO — NH — Drug

II

b) i) synthesis of the reactive polymeric precursor

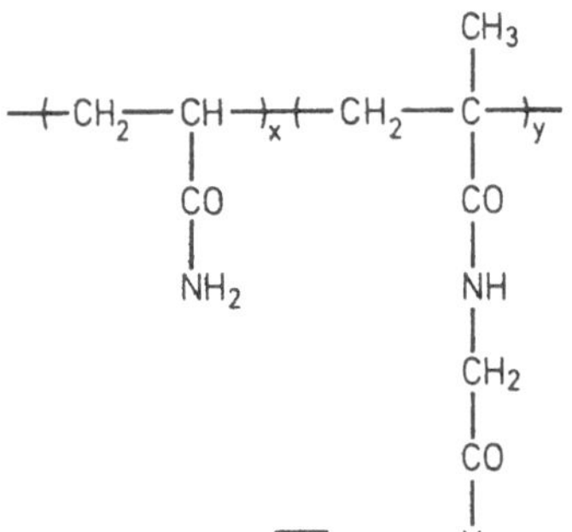

$$+CH_2{-}\underset{\underset{NH_2}{\overset{|}{\overset{CO}{|}}}}{CH}{\to}_x{+}CH_2{-}\underset{\overset{CO}{|}}{\overset{CH_3}{\overset{|}{C}}}{\to}_y$$

with side chain: CO — NH — CH₂ — CO — X

III

X is an electron-withdrawing group, eg.

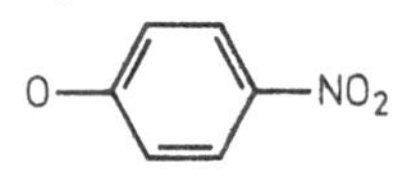

$$O{-}\!\!\!\bigcirc\!\!\!{-}NO_2$$

ii) polymeranalogous binding of the drug

III + H₂N-Drug → II

The following functional moieties are generally used to bind drugs to polymeric carriers:

Ester	—CO—O—
Amide	—CO—N<
Urethane	—O—CO—N<
O-Acylhydroxamic acid	—CO—NH—O—CO—
Hydrazone	—CO—NH—N=C
Thioether	—CH₂—S—CH₂—

In most cases, drugs bound directly to the polymer chain exhibit either a reduced or zero biological activity [70]. For this reason, drugs bound by any of the bonding types listed above should be separated from the polymeric backbone by means of a spacer. Once the drug conjugate reaches the target compartment, the drug can then be split off more readily in its active form. To faciliate the release of the drug it must be attached to the macromolecular backbone by covalent bonds of limited stability in a biological environment [60].

In general, all the bonds mentioned above may be hydrolyzed, although the rate of hydrolysis may vary from very fast to almost insignificant. For instance, the rate of uncatalysed hydrolysis of amide bonds is relatively low but it can be increased by the introduction of specific groups adjacent to the amide bond [71]. Shen and Ryser [72] showed that amides of cis-aconitic acid are cleaved when entering the lysosomal compartment (they are stable at pH 7 and labile at pH 4–5).

$$
\begin{aligned}
&\text{CO} \\
&| \\
&\text{CH}_2 \\
&| \\
&\text{CH} \\
&\quad\searrow \text{COO}^\ominus \\
&\text{CO} \\
&| \\
&\text{NH} \\
&| \\
&\text{Drug}
\end{aligned}
$$

The biological activity of a drug can also depend on its rate of release from the carrier. Molz [70] compared the cytotoxic effect of polymers in which the N,N-bis(2-chloroethyl)amino group was linked to the polymer either via an urethane or O-acyl-hydroxamic bond. The latter polymeric drug was much more effective due to faster release of the active component.

The rate of drug release from the polymeric carrier may be changed within much broader limits if the hydrolysis of the respective bond is catalyzed by enzymes [35, 73–77]. With respect to both synthetic possibilities and the specificity of enzymes present in lysosomes, peptidic, glycosidic or phosphate bonds may be regarded as suitable for the binding of drugs to the polymeric carriers to be used as lysosomotropic drugs [22]. The relationship between the structure of the polymeric drug and its cleavability have been examined in detail for copolymers of N-(2-hydroxypropyl)methacrylamide containing oligopeptidic sequences terminated with p-nitroaniline (NAp) as the drug model.

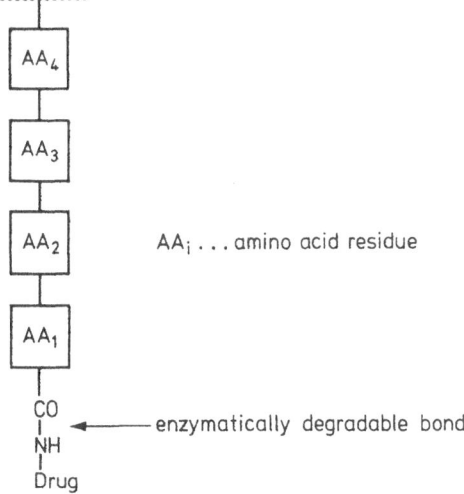

AA$_i$ ... amino acid residue

enzymatically degradable bond

The structure of an oligopeptide spacer can be chosen so as to correspond to the known specificity of certain enzymes [33]. The relationship between the structure of the oligopeptide sequence terminated with NAp and the rate of release of NAp has been determined for chymotrypsin [73], a mixture of lysosomal enzymes isolated from the rat liver [74, 75] and following the pinocytic uptake of yolk sacs cultured in vitro [78]. From the data obtained with lysosomal enzymes it was concluded that the lysosomal thiol-proteinases are particularly important in the degradation of HPMA copolymer side chains. Therefore, the hydrolysis of these polymers by isolated lysosomal thiol-proteinases, cathepsin B [75], cathepsin L [79] and cathepsin H [79] has also been investigated. The knowledge of specific enzyme cleavage of polymers is necessary if one wants to predict the rate of drug release from a known structure of the polymeric carrier or, conversely, if one wishes to suggest the relevant structure necessary for a required rate of drug release.

2.2.3 Examples of Preparation of Polymeric Drugs

2.2.3.1 Preparation of Polymerizable Drug Derivatives

Molz et al. [80] synthesized monomers containing the cytostatic bis(2-chloroethyl)amino group linked via urethane or O-acylated hydroxamic acid bonds to polymerizable methacrylic acid derivatives. Copolymers with hydrophilic monomers, e.g. 2-(methylsulfinyl)ethyl methacrylate, yielded biologically active compounds of the structure

Polymer	X
I	$NH-C_6H_4-O$
II	$O-(CH_2)_2-O$
III	$NH-O$

As already mentioned polymer III was the most active because of the relatively high rate of release of the active compound $HN(CH_2CH_2Cl)_2$.

Hofmann et al. [81] prepared a polymerizable derivative of streptomycine sulfate $(Str-CH=O)$ by reacting the aldehyde group of the drug with methacrylic acid hydrazide

Water-soluble copolymers were prepared by copolymerizing this derivative with methacrylamide or 2-(methylsulfinyl)ethyl methacrylate.

Obereigner et al. [82] synthesized polymerizable derivatives of an antidiabetic drug, N-(4-aminobenzenesulfonyl)-N'-butylurea, where the active substance is bound to a methacrylate moiety via chains of various lengths containing amide or urethane links

$$CH_2\!=\!\overset{\overset{\displaystyle CH_3}{|}}{C}\!-\!X\!-\!NH\!-\!\!\!\bigcirc\!\!\!-\!SO_2\!-\!NH\!-\!CO\!-\!NH\!-\!C_4H_9$$

X...CO

$$CO\!-\!O\!-\!CH_2\!-\!CH_2\!-\!O\!-\!CO$$

$$CO\!-\!NH\!-\!(CH_2)_5\!-\!CO$$

$$CO\!-\!O\!-\!CH_2\!-\!CH_2\!-\!O\!-\!CO\!-\!NH\!-\!\underset{\underset{\displaystyle H_2C\!-\!\bigcirc}{|}}{CH}\!-\!CO$$

$$CO\!-\!NH\!-\!CH_2\!-\!CO\!-\!NH\!-\!\underset{\underset{\displaystyle H_2C\!-\!\bigcirc}{|}}{CH}\!-\!CO$$

A number of examples of the synthesis of polymerizable drug derivatives have been reported which are beyond the scope of this article. However, it should be borne in mind that in most cases, simply the introduction of the polymerizable group into a drug molecule is not sufficient to produce a biologically active drug conjugate. The drug must be separated from the polymer chain by a spacer and practically this may turn out to be a difficult synthetic operation.

2.2.3.2 Polymeranalogous Drug Binding

This procedure may be advantageous since it offers the possibility to introduce greater variability into the polymeric structure, of the carrier. For instance, there are several different synthetic procedures of binding drugs by means of a tetrapeptidic spacer. These are as follows:

From these possibilities the most appropriate procedure can be selected, depending on the relative ease of preparation of the intermediates (crystallizability, yield, accessibility, and the possibility of choosing reaction conditions under which the active substance is not inactivated).

The drug may be linked to the polymeric carrier using a number of reactions with the participation of functional groups, which are either originally present in the polymer of alternatively formed by activation. Another possibility is the use of an activated comonomer in the synthesis of a reactive polymeric precursor. Some typical reactions are described in the following:

1) The use of —COOH groups

$$\left|{-}...{-}COOH \;\; + \; HO{-}Drug \rightarrow \right|{-}...{-}CO{-}O{-}Drug \;\; \left(\text{or} \; \right|{-}...{-}OH + HOOC{-}Drug\,\right)$$

$$\left|{-}...{-}COOH \;\; + \; H_2N{-}Drug \rightarrow \right|{-}...{-}CO{-}NH{-}Drug$$

The functional groups in the scheme are usually joined by means of the carbodiimide method [83] or the method of mixed anhydrides [84]. In these techniques the drug is attached to the polymeric carrier either by an ester or an amide bond.

Chlorambucil [4-[4-/Bis(2-chloroethyl)amino/phenyl)butyric acid] has been bound to vinylpyrrolidone and vinylamine copolymers via an amide bond [59]. The reaction was carried out in chloroform using dicyclohexylcarbodiimide.

When the binding reaction is carried out in aqueous medium, water-soluble carbodi-imides should be used, e.g. [85] N-cyclohexyl-N-[2-(N-methylmorpholino)ethyl]carbo-diimide p-toluenesulfate.

2) Aminolysis of reactive esters

$$\left\{ \cdots - CO - O - \underset{}{\bigcirc} - NO_2 \right. + \ H_2N - Drug \longrightarrow \left\{ \cdots - CO - NH - Drug \right.$$

A number of other electron-withdrawing groups may be used, e.g.

This reaction is very suitable for binding compounds which contain an aliphatic NH_2 group [86]. It was used, for example, in binding deacetylcolchicine to copolymers of N-(2-hydroxypropyl)methacrylamide [79].

3) The use of $-CO-HN-NH_2$ groups (formation of hydrazone)

$$\cdots-CO-NH-NH_2 + Drug-CH=O \longrightarrow \cdots-CO-NH-N=HC-Drug$$

Molz [70] linked daunomycin by means of a hydrazo bond to the copolymer of
N-(2-hydroxypropyl)methacrylamide

It should be noted that daunomycin is a drug which can also be attached to polymeric carriers by means of other procedures. For instance, the amino group of the
amino sugar daunosamin (which is part of the daunomycin molecule) may be used
for this purpose [79].

4) Use of the reactive oxirane ring

This method was used to link a potent beta-adrenergic antagonist Alm-NH$_2$
(which contains a reactive amino group) with the copolymer allyl glycidyl ether/
acrylamide [88]

5) Bond formation with polymers activated by BrCN

This particular reaction is suitable for polymers containing hydroxy groups. Recently, it has been shown [89] that it can be succesfully used to bind chymotrypsinogen to poly[N-(2-hydroxypropyl)methacrylamide], i.e. a polymer which possesses secondary OH groups.

6) Use of anhydride groups which are part of the polymeric backbone

The reaction was used in binding the activated 4-alkylthioderivatives of cyclophosphamide to DIVEMA (divinyl ether and maleic anhydride) copolymer [90].

7) Many other reactions may be applied to bind drugs to polymeric carriers, e.g. the use of polymers activated by cyanuric chloride [91] and the formation of azo bonds [92].

2.3 Incorporation of Residues for Targeting of Polymers

The ultimate goal of binding drugs to carriers is to obtain drugs that recognize and interact specifically with the target cells [93].

Cells display their individuality on their surface by the presence of a number of specific receptors or cell antigens. Advantage could be taken of these identifying marks for selective concentration of drugs by means of appropriate carriers. Let us discuss the binding to polymeric carriers of two specific determinants — namely sugar residues and antibodies. In principle, the binding of targeting moieties may be accomplished by the same binding reactions as those used for the binding of drugs or proteins to synthetic polymeric carriers.

2.3.1 Attachment of Sugar Residues

There are a number of recognition systems [20] in which carbohydrate moieties participate.

How can these units be bound to polymeric carriers? The majority of data reported in the literature relating to sugar binding reactions concerns the modification of proteins (with the formation of the so-called neoglycoproteins). The binding reactions for the modification of proteins can also be applied to synthetic polymers and they are briefly outlined below.

Lee et al. [94] prepared cyanomethyl 1-thioglycosides of D-galactose, D-glucose, 2-acetamido-2-deoxy-D-glucose, and D-mannose. The cyano group in these cyanomethyl thioglycosides was converted to a methyl imidate group to yield 2-imino-2-methoxyethyl-1-thioglycosides (IME-thioglycosides). These IME-thioglycosides readily react with simple amines, amino acids, and proteins in mildly alkaline medium. Thus, these compounds constitute a new group of reagents for attaching sugars to proteins.

A disadvantage of the imidate method is the difficulty in preparing carbohydrate imidates with a low solubility in dry methanol [95]. A new method [95] was therefore proposed which overcomes the above mentioned difficulties; thus, thioglycosides bearing an ω-aldehyde group in the aglycon were prepared. These thioglycosides were attached to proteins by reductive amination using $NaCNBH_3$ as the reducing agent.

N-[S-(β-D-galactopyranosyl)thioacetyl]aminoacetaldehyde

Kawaguchi et al. [96,97] have shown that certain neoglycoproteins are biologically active regardless of the mode of attachment of the sugar residue, or the length of the spacer arm.

Recently, Duncan et al. [98] modified N-(2-hydroxypropyl)methacrylamide copolymers by introducing a small amount of D-galactosamine, D-glucosamine and D-mannosamine residues. Binding was performed by aminolysis of p-nitrophenyl ester groups. Tyrosine residues were also incorporated into

NH—R ... residues of glucosamine, galactosamine, mannosamine

these copolymers so that, after $[^{125}I]$ iodination, their fate could readily be followed in vivo. The biological consequences are described in Section 3.

2.3.2 Attachment of Antibodies

It is possible to use antibodies themselves as drug carriers, i.e. to bind drugs directly to antibodies without using synthetic polymers as intermediates. However, there are some problems in this approach. The extent to which drugs can be directly covalently bound to immunoglobulins is limited due to progressive loss of antibody activity and/or solubility [99,100]. One possible method for overcoming these problems is to use an intermediate carrier [101,102,103]. Thus, poly(L-glutamic acid) (PGA, M_w = 35,000) was first substituted by p-phenylenediamine mustard (PDM) using the carbodiimide method. In this way a PDM–PGA complex in which approximately 16 % of the available carboxy groups had been substituted was obtained. This complex was then combined with rabbit anti EL-4 Ig by a second carbodiimide reaction to give PDM-PGA-Ig in the molar ratio of approximately 90:2:1. This complex retained 66 % of the original antibody activity and was completely water-soluble.

Wilchek [102] described the binding of daunomycin to antibodies also using a polymer as intermediate. In this case, dextrans and polyacrylhydrazide were used as carriers.

Dextran was oxidized with sodium periodate to the corresponding polyaldehyde and daunomycin was coupled with a part of the aldehydes via its amino sugar. This complex was further bound to the lysines of the antibody. This conjugate can either be used directly or stabilized by further reduction with sodium borohydride.

$$HOH_2C \xrightarrow{} \text{—OH} \quad \xrightarrow{IO_4^{\ominus}} \quad HOH_2C \xrightarrow{} \begin{matrix} CHO \\ CHO \end{matrix} \quad \xrightarrow[Ig]{Daunomycin} \quad HOH_2C \xrightarrow{} \begin{matrix} CH_2{=}N{-}Daunomycin \\ CH_2{=}N{-}Ig \end{matrix}$$

$$\xrightarrow{NaBH_4^{\oplus}} \quad HOH_2C \xrightarrow{} \begin{matrix} CH_2{-}NH{-}Daunomycin \\ CH_2{-}NH{-}Ig \end{matrix}$$

Another mode of binding [102] first involved the modification of daunomycin. Periodate cleavage of the $C_3{-}C_4$ bond of the amino sugar residues of the drug yields aldehyde groups. The periodate-oxidized drug was coupled with linear poly(acryl hydrazide) (PAH). The PAH-daunomycin was linked to the antibody by glutaraldehyde.

Generally it may be said that one should choose such binding methods which do not lead to a loss of activity of the antibody, and moreover make possible the release of the active drug at the target site. Also, it should be remembered that binding of antibodies to reactive polymers is, in fact, a reaction of multifunctional compounds, so that one should select such reaction conditions which prevent an extreme rise in molecular weight [104].

3 Biological Properties of Synthetic Polymers

With regard to the biological properties of synthetic polymers, most studies have been designed, using in vivo biological systems, to test the behavior of novel polymeric compounds against existing pharmacological agents with a view to assessing their therapeutic potential. In such experiments, success can only be evaluated in terms of the overall physiological response observed, and it is often difficult to unravel the underlying cellular mechanisms responsible for an effect (or lack of it). As a consequence, when polymeric derivatives have failed to produce the anticipated biological response, and many tests so far have been somewhat disappointing, it has been difficult to determine the particular characteristics of the polymer which prevent beneficial interaction within the biological environment. Only by using simple in vitro systems in parallel with in vivo experiments can we understand more clearly the mechanisms involved in the cellular processing of polymers. Definition of polymer properties such as the biogradability of the polymer itself or of polymer-drug linkages, affinity for the cell surface, mechanism of capture by cells, intracellular fate, biocompatibility, can only help in the design of polymeric drugs and drug carriers which will fulfill their function in biological systems.

3.1 Biodegradability in Relation to Polymer Drug Linkages

When introduced into the body either as solid implants or as soluble macromolecules, many polymers undergo degradation, either to a limited extent giving rise to lower molecular weight products or alternatively when destruction is complete to yield monomeric or atomic constituents which are reutilized in synthetic metabolism or excreted [105]. Degradation occurs as a result of the hostile physiological environment (which is warm, humid and salty) or by enzymatic mechanisms when the polymer is a suitable substrate. The term biodegradability is probably most correctly applied to the latter category whereas the terms erosion or solubilization more appropriately describe the less specific chemically mediated degradation processes.

Degradation of polymers within the organism was first observed in the case of implants and devices where progressive change in the nature of the polymeric material used sometimes caused severe problems such as loss of essential mechanical properties, release of toxic degradation products or initiation of adverse tissue reactions. Although many materials are selected for their ability to resist degradation and are able to retain certain physical properties for long periods of time, recently materials have been developed in which controlled degradation can be used to advantage. An example is the controlled dissolution of sutures and fracture plates [106,107].

In relation to the polymeric drug carrier, degradability is important with respect to polymer-drug linkages and the long-term fate of polymeric carriers.

The latter will be discussed in more detail later in Section 3.4. while in this section the importance of degradation in drug release is described.

As mentioned earlier if a synthetic polymer carries a drug whose pharmacological activity only results from the interaction of the latter with an externally disposed membrane receptor, the drug-polymer linkage simply ensures that the residues vital for receptor binding are freely available and spaced far enough from the carrier molecule to enable unhindered interaction with the receptor. Degradability of the linkage is not a prerequisite for pharmacological action. However, in the development of a targetable lysosomotropic drug delivery system, the drug-carrier linkage is of utmost importance. If the bond is cleaved before the conjugate arrives at the target site, the drug is released and the carrier becomes totally redundant. Linkages must be chosen which are stable during transit, but susceptible to hydrolysis at the target cell in order that the pharmacologically active drug is liberated. All macromolecular carriers ultimately arrive in the lysosomal compartment of the cell subsequent to their pinocytic capture. Lysosomes contain about seventy different hydrolytic enzymes [18] which are capable of digesting all the naturally occurring macromolecules they would normally expect to meet. Drug-polymer linkages should theoretically be designed such that they are susceptible to hydrolysis by lysosomal enzymes and resistant to attack in other body compartments. If this can be realised in practice, an ideal lyso-somotropic delivery system would result.

Oligopeptide sequences can be incorporated into N-(2-hydroxypropyl)methacryl-amide copolymers (for the different methods used see section 2) which serve as po-tential drug attachment/release sites and experiments have been carried out to in-vestigate whether such sequences could be efficiently cleaved by lysosomal enzymes.

N-(2-Hydroxypropyl)methacrylamide copolymers bearing oligopeptide side chains with a terminal p-nitroaniline (NAp) were incubated with purified lysosomal enzymes (Tritosomes) in vitro, and cleavage of these chains was measured by monitoring the release of NAp. It was shown that certain sequences liberate the terminal residue on incubation at pH 5.5 with lysosomal enzymes [108]. Recently, the importance of lyso-somal thiol-proteinases in N-(2-hydroxypropyl)methacrylamide copolymer side chain cleavage was discovered [74]. The synthesis of side-chain amino acid sequences chosen to match the known specificities of certain lysosomal thiol-proteinases re-sulted in a higher initial rate of NAp release and a greater extent of release (up to 50 % of NAp bound for an incubation period of 5 h) [109]. Incubation of substrates in vitro at acid pH with isolated purified lysosomes which have been broken open to release all their enzymes gives an approximation to the physiological environment which those substrates will meet in vivo. This is probably the most suitable system available for preliminary assessment of the degradation of substrates by lysosomes. Certain degradative processes involve more than one enzyme in the liberation of the terminal residue, and the mixed enzyme system provides the opportunity for such a pathway to proceed to completion as it would in vivo. However, when one wishes to identify the individual enzymes responsible for a particular reaction it is more informative to incubate substrates with purified enzymes. On the basis of the data obtained from such investigations tailor-made drug-polymer linkages may be generated which will be cleaved at definable rates by a known lysosomal enzyme. It has already been found that oligopeptide side chains in N-(2-hydroxypropyl)methacrylamide copolymers can be chosen which are hydrolysed to liberate a terminal NAp residue upon incubation with the purified lysosomal thiol-proteinase cathepsin B [75].

Although there have been no other attempts to investigate specifically lysosomal degradation of synthetic polymer drug linkages, generally, there is an increasing awareness of the importance of spacer groups in the attachment and release of pharma-cological agents to and respectively from polymers. Hirano et al. [110] prepared mer-capto derivatives of cyclophosphamide which were bound to divinyl ether and maleic anhydride copolymers (DIVEMA) using different types of spacers. They found that the type of spacer group located between the cyclophosphamide derivative and DIVE-MA is an important factor in determining the rate of hydrolysis measured in phosphate buffer solution at pH 7.0. The longer the alkyl chain of the spacer, the lower the rate of hydrolysis. Usually, for enzymatic cleavage, there is a correlation between the increasing length of the spacer and an increased rate of degradation. Drobník et al. [111] showed that the degradation of oligopeptide side chains in N-(2-hydroxypropyl)-methacrylamide copolymers by chymotrypsin increases with an increasing number of amino acid residues in the oligopeptide sequence. Similarly, the degradation by chymotrypsin of side chains in acrylamide copolymers containing a terminal L-Phe-NAp moiety markedly decreases with decreasing spacing of the -Phe-NAp residue from the polymer backbone [112]. Phe-NAp side chains incorporated into poly[α,β-

(N-2-hydroxyethyl)-DL-aspartamide] have also been shown to be susceptible to hydrolysis by chymotrypsin [113].

Recently, Verlander et al. [114] developed effective spacer groups for the attachment of β-adrenergic drugs to carrier molecules. Catecholamine derivatives were prepared in which the N-isopropyl group was extended by a linear alkyl chain of varying length terminating with a carboxy group. These were then bound to monodisperse peptide carriers containing p-aminophenylalanine as the point of attachment for the drug. Evaluation of the structure — activity relationship showed that all drug derivatives displayed biological activity which was sensitive to structural modification in the spacer far removed from the drug itself.

When the polymeric backbone is susceptible to hydrolysis, drug release may eventually result following degradation of the carrier. Ryser and Shen have shown that methotrexate (MTX) covalently bound to poly(L-lysine) (one molecule of drug per 27 lysyl residues) is able to impair the growth of a normally methotrexate — resistant Chinese hamster ovary cell line [115]. MTX-poly(L-Lys) is regarded as a lysosomotropic drug, and release of reactivated MTX occurs as a result of lysosomal degradation of the carrier. When poly(D-lys) was used as an alternative carrier no inhibition of growth was observed confirming the importance of carrier degradation in the liberation of active drug [116]. The incorporation of a triglycyl spacer between MTX and poly(D-Lys) restored the ability of the conjugate to inhibit the growth of Chinese hamster ovary cells [117]. Chu and Howell [118] have also shown that methotrexate covalently bound to molecular weight fractions of poly(L-lysine) (M = 3,000 or 60,000) displays biological activity.

Polyglutamic acid is another synthetic polypeptide which is used as a drug carrier due to its biodegrability. Van Heeswijk et al. [119] succeeded in binding adriamycin to poly(L-glutamic acid) via N-γ-glutamyl linkages, thus achieving very high degrees of substitution (up to 22 % by weight). In this case, it was intended to release the active drug from the conjugate by γ-glutamyltransferase, an enzyme which is present in high concentrations in certain tumour cells.

It has already been shown that the conjugate is stable in the presence of plasma pro-
teinases and cytocidal towards B-16 melanoma cells. During drug liberation two
degradative processes proceed concurrently, proteolytic hydrolysis of the carrier
and cleavage of the polymer-drug linkages.

Grolleman et al. studied the release of Naproxene and a closely related model
compound (phenylacetic acid) from a degradable polyorganophosphazene conjugate [49].

Several spacers were used and a lysine ethyl ester proved most successful. It was found
that the release of drug is influenced by the degree of substitution and molecular
weight of the polymer. The release measured in rat plasma was approximately twice
that measured in a buffered solution, probably as a result of esterase activity.

Although enzyme specificity, if it could be controlled, would allow the formation
of drug-polymer linkages that are only cleaved at the target cell, other physico-chemical
properties of these linkages may also be important. Due to the fact that the lysosomes
are acidic [23], Shen and Ryser [72] examined the possibility of using this lysosomal
feature to facilitate the release of daunomycin from either Affi-Gel 701 (poly[N-(2-
aminoethyl)acrylamide] beads, Bio-Rad Laboratories, a model system) or poly(D-
Lys). A cis-aconityl linkage between the drug and Affi-Gel 701 is hydrolysed much
more readily at pH 4 and 5 than at pH 6 when incubated in buffered solution for
96 h. At pH 7 the linkage was completely stable throughout the experiment. Although
the Affi-Gel conjugates did not inhibit growth of WEHI-5 cells in culture at neutral
pH, an N-cis-aconityl daunomycin-poly(D-Lys) conjugate caused 90 % growth inhibition. From this it was concluded that whereas the Affi-Gel beads are too large to
penetrate the cells, the poly(D-Lys) conjugate is internalised by pinocytosis and
hydrolysed intralysosomally.

Ferruti et al. [120] prepared polymeric derivatives of daunomycin by reacting its carbonyl group at C-13 or its 3'-amino group with a polymer. In the former reaction
daunomycin is bound to a polymeric hydrazide, poly[N¹-methacryloyl-ε-amino-
caproylhydrazine], and in the latter process daunomycin is reacted with the bis-

succinic half-ester of a poly(ethylene glycol). Although little is known of the rates or mechanisms of the release of drugs from these conjugates, preliminary results have revealed that the hydrazonic derivatives are slightly more active against P388 leukemia when tested in vivo.

3.2 Pinocytic Capture of Polymers by Cells

As stressed earlier, polymers can only enter cells by endocytosis and all polymeric material present in the extracellular environment will be engulfed by the surrounding cells at a rate related to its affinity for the cell surface. Material will only be excluded if the particles or molecules are too large to penetrate vacuoles/vesicles forming at the cell suface. Although the mechanism of pinocytosis, and in many cases its physiological role, is now well understood, most studies have been confined to the examination of the capture of natural macromolecules (proteins, lipids and carbohydrates). Sophisticated in vitro cell culture techniques have been developed which allow us to determine quantitatively the rates of uptake of macromolecules and to ascertain the kinetics of the latter since simple incubation media allow a precise estimation of membrane binding without the interference of competing molecules and with the use of substrates, which are either radiolabelled, labelled by fluorescence or enzymatically active, the rates of uptake can accurately be determined. Such techniques are discussed in more detail in chapter 1. Of course it is difficult to predict from in vitro observations the rates of uptake of polymers under physiological conditions. One reason for this is that the substrate is exposed to a large number of different cell types.

There have been comparatively few attempts to define the kinetics of uptake of synthetic polymers. Some examples are given in Table 1. Uptake of ^{125}I-labelled polyvinylpyrrolidone (PVP) by rat visceral yolk sacs [121], rat peritoneal macrophages [122], and pig aortic smooth muscle cells [123] has been measured in vitro. In each case the mechanism of uptake was that of fluid-phase pinocytosis, the rate of uptake being directly proportional to the concentration of ^{125}I-labelled PVP in the culture medium. Bridges and Woodley also used ^{125}I-labelled PVP as a macromolecular marker to follow pinocytosis in the small intestine of the adult rat [124]. Sections of the small intestine were everted and tied off to form everted sacs which could be maintained in culture for several hours. Uptake of the radiolabelled substrate into the intestinal wall and across it, into the internal sac fluid, was measured. Although it has frequently been reported that macromolecules cannot pass across the intestine, Bridges and Woodley found that ^{125}I-labelled PVP is not only taken up by enterocytes but also passes across to the serosal side, radioactivity appearing in the gut sac fluid only after 15 min of the incubation. More recently, Rowland and Woodley [125] reported that uptake of ^{125}I-labelled PVP by rat intestinal sacs is proportional to concentration and may be inhibited by microtubule and metabolic poisons. It was concluded that ^{125}I-labelled PVP is transported into the gut by fluid-phase pinocytosis and presumably across it by subsequent exocytosis, transport across the cells being described by the term diacytosis. They also demonstrated that entrapment of ^{125}I-labelled PVP in liposomes increases its rate of capture by the gut.

Copolymers of N-(2-hydroxypropyl)methacrylamide whose oligopeptide side chains contain [^{125}I]iodotyrosine are captured by the rat visceral yolk sac at a rate

Table 1. Pinocytic uptake of some synthetic polymers by cells cultured in vitro

Polymer	Cell Type	Mechanism of Pinocytosis	Rate of Pinocytosis	Ref.
	rat visceral yolk sac	fluid-phase	1.71[a]	[121]
	rat peritoneal macrophage	fluid-phase	0.034[b]	[122]
polyvinylpyrrolidone	pig aortic smooth muscle	fluid-phase	0.123[a]	[123]
	pig endothelial cells	fluid-phase	0.034[a]	[123]
	rat small intestine	probably fluid-phase	0.96[a]	[124]
DIVEMA	rat peritoneal macrophages	adsorptive	3.30[b]	[127]
	mouse peritoneal macrophages	—	—	[128]
	rat peritoneal macrophages	probably adsorptive	—	[129,132]
poly(L-lysine)	sarcoma 180	probably adsorptive	—	[130]
	mouse fibroblasts	probably adsorptive	—	[131]
poly(D-lysine)	sarcoma 180	probably adsorptive	—	[130]
poly(L-ornithine)	sarcoma 180	probably adsorptive	—	[130]
poly(9-vinyladenine)	3T3 cells	probably adsorptive	—	[134]
poly(l-vinyluracil)	3T3 cells	probably adsorptive	—	[134]
poly[α,β-(N-2-hydroxyethyl)-DL-aspartamide]	rat visceral yolk sacs	adsorptive	10[a]	[62]
poly[vinylpyrrolidone-vinylamine]copolymers	rat visceral yolk sacs	adsorptive	2.75–8.15[a]	[136]
	rat peritoneal macrophages	adsorptive	0.110–0.689[b]	[136]
poly[N-(2-hydroxypropyl)methacrylamide] copolymers	rat visceral yolk sac	fluid-phase	2[a]	[78]

* Data are expressed in clearance terms that is the volume of culture medium (μl) whose substrate is captured per mg protein (a) or per 10^6 cells (b) per hour of incubation

consistent with uptake being by fluid-phase pinocytosis [78]. In the latter experiments the radiolabel may theoretically be liberated from the side chains within the cell, depending on their susceptibility to lysosomal hydrolysis. Certain side chain sequences (-Gly-Gly-[125I]Tyr-NAp or -Gly-Phe-[125I]Tyr-NAp) are readily cleaved to release 125I-iodotyrosine (detectable in the culture medium using Sephadex G-15 column chromatography) whilst others were completely resistant to hydrolysis (-Gly-β-Ala-[125I]Tyr-NAp). Such data illustrate the potential of polymer-drug linkages for rate-controlled intracellular drug delivery. When a proposed drug carrier is itself internalised by fluid-phase pinocytosis it presents the maximum opportunity for incorporation of additional residues into the polymer, which will target the carrier to a specific cell type. When the carrier itself has a high affinity for cell membranes, this is more difficult to achieve.

In recent years, reports describing the adsorptive pinocytosis of synthetic polymers by a variety of different cell types have appeared. The polyanion DIVEMA has aroused much interest due to the wide variety of different biological activities it displays [126] including antineoplastic activity and the ability to activate macrophages. The rate of pinocytic capture of [14C]DIVEMA and 125I-labelled DIVEMA (a p-methoxyaniline derivative of DIVEMA) was measured with a view to elucidating the mechanism of DIVEMA action. Capture of labelled DIVEMA by rat yolk sacs is only about two times faster than that of a fluid-phase substrate, whereas capture by rat-peritoneal macrophages is about hundred times faster [127]. Uptake of DIVEMA by macrophages may be attributed to a highly efficient receptor-mediated process and may be related to the ability of DIVEMA to "activate" macrophages. It has also been reported [128] that mouse peritoneal macrophages accumulate [14C]DIVEMA but no attempts have been made so far to measure the rate or define the kinetics of capture.

Several years ago Seljelid et al. [129] showed that poly(L-Lys) enhanced the uptake of 3H-labelled reovirus double-stranded RNA by macrophages and similarly it has been shown that poly(L-Lys), poly(D-Lys) and poly(L-ornithine) markedly increase the rate of uptake of 131I-labelled albumin by Sarcoma S-180 cells grown in monolayer [130]. In the latter experiments, Ryser found that the ability to stimulate uptake is clearly related to the molecular size of the polymer used. Recently, a horseradish peroxidase (HRP)-poly(L-Lys) conjugate was prepared and used to determine uptake by L-929 mouse fibroblasts grown as monolayers [131]. Measurements of the enzymatic activity of HRP associated with cells exposed to either HRP alone (150 μg/ml) or the HRP-poly(L-Lys) conjugate (15 μg/ml) have shown that intracellular levels of HRP detectable in the presence of conjugate are about 43 times higher than those detected after exposure to the higher concentration of free HRP. A [3H]methotrexate-poly(L-Lys) conjugate was also used to follow the uptake of the conjugate by both normal and methotrexate transport-deficient Chinese hamster ovary cells [115]. It was shown that the conjugate enters both cell types very rapidly and that in the transport-deficient cells uptake of the polymer bound drug is 200times greater (after 60 min exposure) than that of the free drug. In transport-proficient cells uptake of the conjugate was only about 40 times greater at this time. Precise quantitation of the rates of capture of poly(L-lysine) and poly(L-orthinine) is difficult due to their high affinity for the cell surface. When both are labelled with [125I]iodide and added to cultures of rat yolk sacs or rat peritoneal macrophages, it was found that whereas high levels of radioactivity are detectable in both cell types the addition of inhibitors to the incubation media known to inhibit pinocytic uptake did not prevent cell association from radioactivity [132]. It was concluded that a large portion of radioactivity represented material bound to the external surfaces of the cells or the tissue and not material internalised by pinocytosis.

Chu and Howell [133] have measured the uptake of [^{3}H]methotrexate poly(L-lysine) conjugates ($\overline{M}_w$ = 3,000 or 60,000) by W 1-L2, T24A and human bone marrow cells in vitro. Both conjugates were taken up more rapidly than free [^{3}H]MTX by all three cell lines, the higher molecular weight derivative being captured the most rapidly. The uptake of the conjugate by T24A tumour cells was significantly higher than that by W1-L2 or bone marrow cells. This observation is in agreement with the IC$_{50}$ of the MTX-poly(L-Lys) derivatives against these cells. Uptake of MTX-poly(L-Lys) is only inhibited to approximately 50% of the control value when the incubation temperature is lowered from 37 to 0 °C (conditions under which pinocytic uptake is normally inhibited). Lysis of W1-L2 or T24A cells has confirmed that 65–75% of radiolabel is always associated with the membrane fraction even after incubation at 37 °C. This confirms that most of the labelled material is externally disposed.

Poly(9-vinyladenine) and poly(1-vinyluracil) are electroneutral polymers which are believed to be stable to enzymatic hydrolysis. They form complexes with complementary polynucleotides and display antiviral activity. Noronha-Blob et al. [134] measured the uptake of [^{14}C]poly(vU) and [^{14}C]poly(vA) by 3T3 cells grown in culture. They found that uptake depends on the concentration of both polymers and that [^{14}C]poly(vU) is taken up more rapidly than [^{14}C]poly(vA), some ten times more [^{14}C]poly(vU) being associated with the cells after 90 min. By incubating cells in a medium containing [^{14}C]poly(vA) for 48–72 h and then subculturing through several generations in a medium lacking the radioactive substrate, it was shown that poly(vA) is retained in the cells from one generation to the next. Cell fractionation showed that about 50% of the radioactivity from [^{14}C]poly(vA) is present in the lysosomal-mitochondrial fraction, the rest being found in the soluble fraction. This may be due to either transfer of poly(vA) into the cytoplasm via the plasma membrane or lysosomal membrane or alternatively to the disruption of lysosomes during the fractionation procedure. In either case, a considerable portion of the accumulated polymer is captured by adsorptive endocytosis and concentrated intralysosomally.

There is a great need to identify and characterise the sites in the structure of macromolecules which are responsible for their affinities for cell membranes. Once this information becomes available it will be possible to synthesize tailor-made polymers with a membrane affinity meeting the desired functional requirements. An interesting approach which has been used to define polymer-membrane interactions is that of differential scanning calorimetry (DSC) of model membrane systems. The heat capacity profiles obtained by DSC show a series of peaks which can be related to domains in the membrane and hence any interaction with a polymeric solute that causes structural changes in the membrane can be measured. Tirrell and Boyd [135] have used this technique, coupled with cell agglutination measurements and direct binding measurements with radiolabelled polymers, to investigate interactions of polymers with the cell surface. Human red blood cells or erythrocyte ghosts were used as model membranes and a large number of polymers were tested including poly[N-(2-hydroxypropyl)methacrylamide], polyvinylpyrrolidone, polyacrylamide, poly(acrylic acid), and poly(ethylene oxide). It was found that all neutral polymers, poly[N-(2-hydroxypropyl)methacrylamide], polyvinylpyrrolidone and poly(ethylene oxide), cause no pertubation of the heat capacity profile. This observation is consistent with the data reported earlier (see pp. 40) which indicate that polymers of this type do not bind to the membranes of cells in culture. Poly(methacrylic acid) causes slight but significant changes of the heat capacity profile whereas poly(acrylic acid) causes no changes indicating the perturbative effect of the hydrophobic α-methyl group. The polycation "polybrene" (poly(dimethyliminiotrimethylenedimethyliminiohexamethylene dibromide)) strongly interacts with the negatively charged spectrin molecule in the erythrocyte membrane.

Another approach concerned with polymer-membrane interactions is the use of cells in culture to examine the effects of structural modifications of polymers on their rate of pinocytic capture. Incorporation of approximately 20% tyramine residues into poly[(α,β-(N-2-hydroxyethyl)-DL-aspartamide] greatly increases the rate of uptake of the polymer by rat visceral yolk sacs incubated in vitro [62]. Similarly, an increase of the proportion of oligopeptide side chains bearing a terminal tyrosinamide in N-(2-hydroxypropyl)methacrylamide copolymers causes an increase in their rate of pinocytic capture by the same tissue [63]. Both processes are thought to be mediated by the increased hydrophobicity of the substituted copolymer. Pratten et al. [136] used vinylpyrrolidone-vinylamine copolymers to study the effect of cationic charges on the uptake of a synthetic polymer. Copolymers with an average molecular weight of 46,000 or 120,000 containing vinylpyrrolidone and vinylamine in the ratio 10:0.77 were prepared. These copolymers were labelled using either the Bolton-Hunter reagent or methyl 3,5-di[^{125}I]iodohydroxybenzimidate which, in contrast to the Bolton-Hunter reagent, preserves the same overall positive charge or the labelled copolymer molecule. These labelled copolymers were used as substrates for the pinocytic uptake by either rat visceral yolk sacs or rat peritoneal macrophages cultured in vitro. The copolymer derivative with $M_w = 46,000$ and labelled with the Bolton-Hunter reagent is captured two times more rapidly than the rate of uptake of a substrate captured in the fluid-phase by yolk sacs and three times more rapidly by macrophages. The higher molecular weight derivative is captured more rapidly by macrophages, the fastest rate of uptake was measured after benzimidate labelling and was about twenty times faster than the rate of uptake of a substrate captured in the fluid-phase.

It has been demonstrated that the rate of pinocytosis of an adsorptive substrate is related to its molecular weight. Even when the densities of the binding sites on macromolecules of different size are the same, the higher molecular weight polymer is bound more strongly and hence captured more rapidly. Different molecular weight fractions of ^{125}I-labelled polyvinylpyrollidone were used to study the effect of molecular weight on the rate of uptake of this substrate which is captured by fluid-phase pinocytosis [29]. The rate of pinocytic uptake by rat visceral yolk sacs has been found to decrease with increasing molecular weight of the substrate. Conversely, when using rat peritoneal macrophages as a model system the higher molecular weight material is captured more rapidly. Chromatographic analysis of ^{125}I-labelled PVP actually internalised by the two cell types confirmed that the yolk sac selects low molecular weight molecules from the range presented to it whereas the macrophage is less discriminating. The observation that high molecular weight material is not intenalised by the yolk sac is not surprising in view of the ultrastructure of this tissue. The apical region of the cell is microvillous in nature and the newly forming pinocytic vesicles originate at the base of

these microvilli. This area acts as a barrier or sieve limiting access to the forming pinocytic vesicles to the smaller molecules ($M < 10^5$). A similar effect is observed when a
copolymer of N-(2-hydroxypropyl)-methacrylamide whose chains are crosslinked by
the sequence Gly-Gly-Phe-Tyr-HMDA-Tyr-Phe-Gly-Gly is fractionated, the fractions labelled and subsequently used as a substrates for pinocytosis in the rat visceral
yolk sac [137]. Only the fraction with a mean molecular weight of 3.4×10^4 is taken up
at the expected rate of fluid-phase pinocytosis. Fractions with higher mean molecular
weights ($\overline{M}_w = 1.1 \times 10^5, 1.55 \times 10^5, 4 \times 10^5$) are taken up more slowly, the rate of
capture being proportional to the mean molecular weight of the fraction.

Since the ability of cells of the reticuloendothelial system to phagocytose high
molecular weight molecules and particulate matter is well-known, we only refer to
a recent study of Fornůsek et al. [138] who have used methacrylate copolymer particles
(mean diameter 0.5 μm) to distinguish between normal and stimulated macrophages.

That cells discriminate between pinocytic substrates according to size, even at relatively low molecular weight of the substrate is very important in relation to the idealised
drug carrier. Obviously, a polymeric carrier which is too large to penetrate into the
pinocytic vesicles of the desired target cell must not be chosen. Alternatively, one
might purposely select a "larger" vehicle if the desired cell type is able to engulf such
material, especially when doing so aids in targeting synthetic polymers preventing the
drug conjugate from being absorbed by other cell types.

From the preceding considerations it is obvious that the pinocytic uptake of a polymer by cells is affected by fundamental parameters such as overall charge, hydrophobicity, and not least molecular weight. These parameters are crucial in the design
of efficient drug delivery systems. Much more work is still required in this area to
elucidate the mechanism of capture of the polymers which are believed to be valuable
drug delivery systems. In the necessary experiments several factors require special
consideration, for example:

1) Polymers with unknown molecular weight distribution or high polydispersity
   should not be used. It is becoming increasingly obvious that the molecular weight
   of the polymer is of paramount importance in determining its biological behaviour.
   Well characterized preparations should always be used.
2) Progressive accumulation of radioactivity by cells is not always indicative of pinocytic uptake and efforts should be made to distinguish between extracellular
   binding and true pinocytic internalization.
3) Progressive accumulation of radioactivity by cells is not a reflection of the rate
   of pinocytic capture if the substrate is digested intracellularly, thus releasing low
   molecular weight labelled products which can diffuse back out of the cell.
4) When examining the effect of polymer substituents on membrane binding and
   hence rate of uptake, efforts should be made to limit the number of variables, i.e.
   by using polymers with the same molecular weight distribution, but different
   degrees of substitution.

3.3 In Vivo Targeting of Polymers

Although in vitro systems give a useful insight into the ways that cells capture polymers, for purposes of drug delivery it is essential to demonstrate in vivo that polymers

can be directed to specific cell types. So far there are only very few data available on this particular problem.

The blood clearance and distribution of polyvinylpyrrolidone in the body after intravenous administration have been described in several articles. Originally, interest stemmed from the clinical use of PVP as a plasma expander and the "storage disease" effects which were subsequently reported. Recently, this compound has been used in vivo as an investigative tool due to its capture by cells fluid-phase pinocytosis. Ravin et al. [139] used ^{131}I-labelled and [^{14}C]PVP to determine whether or not PVP can be metabolised by the body and examined the blood clearance characteristics and the long-term fate of the polymer. When ^{131}I-labelled PVP fractions with a mean molecular weight of 2.8 to 5.0×10^4 are injected intravenously into rats, the rate of blood clearance has been found to be related to the size of the PVP used. The rate of clearance also correlates well with the appearance of radioactivity in the urine and the threshold for the filtration of PVP by the kidney is assumed to be at molecular weights between 3.5 and 4.0×10^4. Regoeczi [140] used ^{131}I-labelled PVP to monitor the activity of reticuloendothelial cells in rabbits and showed that after in vivo administration of ^{131}I-labelled PVP, radioactivity disappeares slowly from the circulation due to its relatively low affinity for phagocytic cells. This may be useful for measuring the endocytic activity in pathological states.

Munniksma et al. [141] extensively studied the mechanism of the in vivo capture of ^{125}I-labelled PVP. They confirmed that capture occurs by fluid-phase pinocytosis and measured the rates of pinocytosis by many cell types. The results they obtained on the blood clearance of ^{125}I-labelled PVP using normal and nephrectomised rats again underline the importance of glomerular filtration in this process.

Poly[N-(2-hydroxypropyl)methacrylamide] is another polymer developed as a prospective plasma expander and as such its rate of elimination from the bloodstream was of interest. As mentioned in section 2, experiments were carried out using [^{14}C]polymer [28] and the results obtained show that clearance is clearly related to the molecular weight of the polymer. Recently, by using N-(2-hydroxypropyl)-methacrylamide copolymers whose tyrosine-containing side chains were labelled with [^{125}I]iodide it has been shown [98] that the pattern of blood clearance from rats is consistent with that reported for a similar molecular weight distribution of ^{125}I-labelled PVP, i.e. these data show that N-(2-hydroxypropyl)methacrylamide copolymers are captured, like polyvinylpyrrolidone, non-selectively in vivo by fluid-phase pinocytosis.

As stressed earlier, the importance of carbohydrate residues in pinocytic recognition systems has become increasingly obvious over the years (reviewed in Ref. [20]). Serum glycoproteins bearing exposed terminal galactose residues are rapidly and specifically cleared by liver hepatocytes [142]. Cells of the reticuloendothelial system (e.g. liver kupffer cells, alveolar and peritoneal macrophages) contain receptors which recognise mannose, glucose, and N-acetylglucosamine residues [143,144]. In contrast, fibroblast membranes contain a receptor which avidly binds mannose-6-phosphate and is functional in implementing the rapid uptake of certain lysosomal enzymes [145]. It is now becoming clear that these systems may be used to enable targeting synthetic polymers. The incorporation of 2 to 4 galactose residues into N-(2-hydroxypropyl)-methacrylamide copolymers ($M_w \sim 5.0 \times 10^4$) produces a blood clearance in the rat which is much faster that of the unmodified copolymer. This may be attributed to the rapid pinocytic capture of the copolymer by the liver [98] and impressively shows that a synthetic polymer can be successfully directed to a target organ.

Abel et al. [146] prepared poly{N-[4-(2-pyrimidinylsulfamoyl)phenyl]acrylamide} because the residue N-pyrimidin-2-yl-4-aminobenzenesulfonamide has been shown to concentrate selectively in an animal tumour. They studied the tissue distribution of this [^{14}C]labelled polymer in Balb/C mice bearing a subcutaneously transplanted PC6 plasmacytoma. In these experiments the polymer was injected intraperitoneally and the uptake of radioactivity by liver and PC6 tumor followed. The amount of radioactivity found in the liver was very high in comparison with that in the tumor and probably reflected the non-specific concentration of macromolecules in liver which often results after intraperitoneal administration.

Renal filtration and pinocytic capture are two processes which effect the distribution of a polymer in the body. There are many processes such as penetration through capillary walls about which we have only little or no information at all. Some macromolecules may be transported from one cell type to another, either by direct transfer or exocytic release followed by pinocytic uptake by a different cell type. It has been shown that polymers such as PVP which can persist in cells for long periods are also slowly excreted by the body. ^{125}I-labelled PVP captured by liver is slowly transported via the bile canaliculi and gall bladder to the intestine and excreted in the faeces [141].

In the physiological environment polymers are free to interact with all the other molecules in the surroundings. Of particular importance is the interaction of polymers with serum constituents such as serum proteins. If polymers are designed with a view to targeting and they interact spontaneously with serum proteins, the observed distribution of these polymers in the body may be radically changed. Ohno et al. [147] studied the interactions between many different polymers and human serum albumin or human serum γ-globulin. They used precipitation technique to assess complex formation and when no precipitate formed, circular dichroism spectra were measured. It has been shown that both anionic and nonionic polymers interact with these proteins. In both cases, pH affects complex formation. Hydrophobic interactions are particularly important during complex formation of nonionic polymers with proteins.

3.4 Long-Term Fate of Synthetic Polymers

The clinical usefulness of a polymer in the treatment of diseases can be severely limited if it persists in the body for a long period after administration, especially if it causes adverse reactions after a period of time. Two main factors influence the rate of elimination of macromolecules from the body. Firstly, the size (molecular weight and molecular conformation) which, as has been discussed in the previous section, controls the potential for a loss of the molecule via the kidney. Secondly, biodegradability ultimately determines the long-term fate of the synthetic polymer. With respect to polymeric drug carriers these factors are interesting points for discussion. Obviously, the ideal carrier is sufficiently large to be retained in circulation (or other body fluids) long enough to permit efficient delivery of drug to the target tissue. However, after completion of this task, ideally, the polymer should be eliminated. Although macromolecules such as poly(L-α-amino acids), which readily undergo degradation, present no problems, they are often immunogenic which is a disadvantage. The synthesis of novel compounds of this type is attracting increasing interest. Recently, Lenz and Guerin have described the synthesis of potentially degradable polymers of malic and aspartic acid [148]

and Braswell et al. the synthesis of poly(α-amino acids) which may also undergo biodegradation [149].

Although the ideal drug carrier may ultimately prove to be a macromolecule susceptible to degradation, a non-degradable molecule may be preferred if it displays other more important beneficial properties such as good biocompatibility. The use of soluble crosslinked polymers is one way to overcome problems of polymer accumulation, and theoretically they could be used as follows. Polymers are chosen whose molecular weight is sufficiently low to allow their passage through the kidney glomerulus. The polymer chains are then crosslinked (for specific examples see below) to a level below the gel point, producing molecules whose size is greater than the renal threshold. The structure of the crosslinks can be chosen so that they are hydrolysed in the body (ideally in the lysosomes), and the lower molecular weight chains then released can be excreted. The feasibility of such a scheme has been examined using N-(2-hydroxypropyl)methacrylamide copolymers crosslinked by oligopeptide sequences. The amino acid sequences of such crosslinks can be chosen such that they are cleaved by the purifield lysosomal enzyme cathepsin B [75] or following the pinocytic capture of crosslinked polymers by cells in culture [137]. Preliminary in vivo experiments have revealed that after intravenous administration of such crosslinked N-(2-hydroxypropyl)methacrylamide copolymers to rats it is possible to detect polymer chains in the urine which have a molecular weight similar to that of the aminolysed precursor [150], thus indicating that the crosslinked copolymer has been hydrolysed in vivo.

A similar concept has been used to develop a biodegradable, water-soluble blood plasma substitute. Fauvarque and Malinge [151] prepared a polymer where poly(ethylene oxide) chains are linked by cleavable pyrophosphate units. They first quantitatively phosphorylated poly(ethylene oxide) 4,000 and then condensated this prepolymer by building the pyrophosphate links.

It has been shown that this polymer is slowly degraded at 37 °C in aqueous buffered solutions (pH 7.3). However, its degradation by phosphatases has not been studied so far.

3.5 Biocompatibility

Biocompatibility is a term used to describe the relationship between a foreign polymeric material and the physiological environment. Since the aspects of blood biocompatibility have been reviewed in detail [152] we will only briefly mention those factors pertaining directly to the drug carrier.

The blood biocompatibility of a drug carrier is obviously very important if it is administered intravenously. However, the drug carrier is only transiently present in the circulation. Therefore, unless the polymer is highly reactive one would not expect to encounter problems such as those reported for materials and devices being constantly or repeatedly in contact with blood.

The question of blood biocompatibility may not even arise when the drug carrier is administered orally, intramuscularly or intraperitoneally. Whatever the route of administration, the immunogenicity of the proposed drug carriers is very important and only very few experimental data on this problem have been reported so far. It has been shown that plasma expanders such as polyvinylpyrrolidone and poly[N-(2-hydroxypropyl)methacrylamide] do not initiate an immune response. However, when such macromolecules are modified either by conversion into copolymers or attachment of pendant pharmacological agents the immunological properties may be totally different [87]. Conjugation of two or more compounds which are not themselves immunogenic may result in the so-called hapten effect. However, it has been shown that binding of proteins to water-soluble polymers may significantly decrease the antigenic properties of the protein. Abuchowski et al. [153] covalently bound liver catalase to poly(methoxyethylene glycols) (PEG) (1,900 daltons) using 2,4,6-trichloro-s-triazine as a coupling reagent and showed that the enzyme was no longer immunogenic. Similarly through attachment of uricase and arginase [154] to PEG conjugates are obtained which retain their enzymatic activity without any immunogenicity.

The drug carrier or any metabolite it might yield should not be carcinogenic, pyrogenic or toxic.

3.6 Pharmacological Activity of Polymeric Drug Carriers

Although there has been some remarkable clinical progress in polymer-based sustained drug release systems an equivalent success for a water-soluble, targetable polymeric drug-delivery system has not been reported so far. With further efforts based on rational design it is surely only a matter of time before an effective targetable system will be found. At present, there are sufficient reports in the literature describing the successful treatment of animal model diseases to give cause for optimism. Some examples of polymeric drug delivery systems are given in Table 2.

Several studies are concerned with the use of poly(L-lysine) as a carrier. Ryser and Shen [115] have shown that a methotrexate-poly(L-Lys) conjugate administered in vitro to a methotrexate-resistant Chinese hamster ovary cell line is cytostatic and hence able to overcome drug resistance. Again using methotrexate linked to poly-(L-Lys), Chu and Howell [138] have found a measure of the differential toxicity towards five cell lines from human solid tumours. Five lines of human lymphocytes are affected to a lesser extent by the conjugate, and the magnitude of differential toxicity is such that a constant infusion of MTX-poly(L-Lys) over long periods may theoretically maintain a serum concentration which is sufficiently high to kill tumour cells without affecting bone marrow cells. Leucovorin competes with MTX for transport across normal and malignant cell membranes [155] and can be used to overcome problems associated with MTX toxicity. Chu and Howell [156] have recently examined the possibility of using a macromolecular carrier, poly(L-Lys), in combination with

Table 2. Some soluble synthetic polymers which have been used as carriers of therapeutic agents

Polymer	Drug	Method of Evaluation	Ref.
DIVEMA	Methotrexate	L1210 leukemic cells and Lewis lung carcinoma in vivo	[158]
DIVEMA	Methotrexate	L5178Y cells cultured in vitro	[157]
DIVEMA	Cyclophosphamide derivatives	L1210-bearing mice	[110]
poly(ethylen-imine)	Methotrexate	L5178Y cells cultured in vitro	[157]
poly(L-glutamic acid)	Adriamycin	L 1210 cells and B-16 melanoma cultured in vitro	[119]
poly(L-lysine)	Methotrexate	methotrexate-resistant Chinese hamster ovary cells in vitro	[115]
poly(L-lysine)	Methotrexate	human solid tumour cells and lymphocytes in vitro	[138]
poly(L-lysine)	Methotrexate	L1210 cells in the pleural cavity	[156]

leucovorin to increase the therapeutic efficiency of MTX against L1210 tumour in the pleural cavity of mice. It was hoped that the carrier would direct the drug action to the tumour cells in the pleural cavity and that systemic administration of leucovorin could be used to neutralise MTX leaking out of this compartment. Although coupling of MTX to the high molecular weight carrier increases its toxicity (giving an increased life span) this may be attributed to the toxicity of the poly(L-Lys) carrier. The toxicity of the carrier itself has proved to be a distinct disadvantage since it limits the amount of drug that can be delivered. Intracavity chemotherapy in conjunction with a systemic neutralising agent does have potential if a carrier can be found which will improve the uptake of the drug into malignant cells.

Fung et al. [157] prepared four synthetic polymeric derivatives of MTX and tested their ability to inhibit the growth of L5178y leukemic cells cultured in vitro. Although DIVEMA-MTX and polyethylenimine-MTX exhibit the same, or a slightly weaker effectiveness than an equal concentration of free MTX, poly(L-Lys)-MTX and carboxymethylcellulose-MTX were considerably more effective. A DIVEMA-MTX conjugate has also been tested against L1210 leukemia cells in mice [158] and the conjugate was found to be more effective than either MTX or a combination of DIVEMA and MTX in producing increased life span.

The antitumour drug cyclophosphamide requires activation by microsomal enzymes before it displays the alkylating activity essential for the pharmacological effect. The active metabolite of cyclophosphamide is 4-hydroxycyclophosphamide which is not very stable and therefore unsuitable for attachment to carriers. Hirano et al. [90] attached stable mercapto derivaties of 4-hydroxycyclophosphamide to DIVEMA and later evaluated the rate of hydrolysis of the polymeric derivatives and their activity against L1210-bearing mice [110]. Unfortunately, the polymeric derivatives of cyclophosphamide are less effective than low molecular weight derivatives in prolonging the life span. This appears to be due to their higher relative toxicity which limits the maximum dose. Recently, the synthesis of micell-forming block copolymers (comprised of polyethylenimine and palmitic acid) containing cyclophosphamide derivatives in the hydrophobic core has been reported [159]. Although it has been shown that drug is released under physiological conditions in vitro the pharmacological activity has not been tested so far.

Hurwitz et al. [160] have bound daunomycin via its carbonyl group to macromolecular hydrazides including a derivative of poly(glutamic acid). In this form, daunomycin is slightly less cytotoxic to mouse lymphoma cells in vitro but seems to be more effective against the same lymphoma in vivo. In order to test and synthesize site-directed drugs, daunomycin-containing macromolecules have also been attached to the γ-globulin fractions of serum from goats immunized with the membranes of mouse lymphoma cells. Although the products contain polymer-bound drugs they tend to precipitate from solutions after 2–12 h; this being a difficult problem to overcome.

Polyvinylpyrrolidone has been suggested as a soluble carrier of proteins [161]. In a model system trypsin bound to PVP has been shown to retain its activity against benzoyl-arginine-p-nitroanilide (98 % active in comparison to an equivalent amount of free trypsin). This system also exhibits a significantly lower degree of self-inactivation than free trypsin. Non-covalent complexation of oxytetracycline with polyvinylpyrrolidone has been used to alter the distribution of the antibiotic in the body after intravenous administration to rats [162]. When injected in a formulation containing polyvinylpyrrolidone a slower accumulation of drug in some organs occurs compared with the distribution observed in the absence of PVP. The distribution of the oxytetracycline in the body observed over a longer period is indicative of an equilibrium between free and bound oxytetracycline which alters with time.

4 Synthetic Polymers Used as Drug Carriers

4.1 Comparison with other Drug Carrier Systems

Other systems which are currently studied as potential drug delivery systems are classified into two groups: Firstly carriers which, by nature, form some kind of multi- or unilamellar vesicle, e.g. liposomes, erythrocyte ghosts, synthetic microcapsules, and secondly carriers based on naturally occurring macromolecules such as albumin, DNA, antibodies, Dextran (reviewed in Ref. [163]).

In recent years a wealth of literature on the use of liposomes as drug carriers has accumulated, and this topic has been extensively reviewed [164, 165]. Although at one time liposomes were thought to be the key to carrier-mediated delivery, largely due to early observations relating to their potential for oral administration of insulin, data collected more recently has revealed some of their drawbacks. Any drug carrier must be stable on its way to the target site, and many types of liposomes do not fulfil this requirement when administered either orally or intravenously. However, it is possible to enrich liposomes with cholesterol and they appear to be more stable in the blood circulation [166]. Liposomes can also be coated with carbohydrates such as those present in bacterial cell walls; this leads to increased stability [167]. In order to prepare stable cell models, Gros, Ringsdorf and Schupp [39] have recently reported the synthesis of liposome-like synthetic polyamphiphiles. These compounds are completely stable under virtually any conditions and studies are now being made to try and incorporate domains into their structure which may be physiologically useful to allow a "cork-like" opening of the capsule at the appropriate location. Couvreur et al. [168] have also described the preparation of so-called "nanocapsules" by micellar polymerization of polyacrylamide. It has been shown that such nanocapsules can be used to increase the uptake of fluorescein into the lysosomes of cultured fibroblasts. Targeting of liposomes to cells other than those of the reticuloendothelial system has also proved difficult to achieve. As a consequence of their size liposomes and also the other carriers of this group are avidly phagocytosed by macrophages and largely excluded from other cell types. Although being disadvantageous in the direct treatment of many diseases including neoplastic diseases, targeting to the reticuloendothelial system can be utilized to treat macrophage-related disorders and also to activate macrophages so that they become tumoricidal [169].

Studies on the toxicity of liposomes have shown that intravenous administration of egg phosphatidylcholine, cholesterol or similar lipids produce no apparent toxicity or histological changes in rats [170], tumour-bearing patients [171] or patients with storage diseases [172]. However, intraperitoneal injection of such liposomes over long periods can cause leucopenia, pulmonary leucopenia and pulmonary leucostasis [173], and intravenous injection of other types of liposomes can result in a variety of undesirable physiological effects including increased catecholamine metabolism [174], affected pituitary function [175], and blood clotting [176]. In many cases, liposomes act as immunological adjuvants [177] and this may be undesirable, especially if they are given repeatedly over a long period.

Trouet and coworkers have extensively studied drug delivery systems. Although they have tested a variety of different carrier vehicles [178] much of their work is concerned with the use of natural macromolecules as drug carriers. In early studies it was found that complexes of daunomycin and adriamycin non-covalently linked to DNA display increased activity compared with free daunomycin against experimental leukemia [179]. Later clinical trials showed a therapeutic effect at least equal to that of the free drug with reduced cardiotoxicity [180]. It was quickly realised that such non-covalent complexes are unsuitable as carriers due to the lack of stability in the bloodstream. Therefore recent work has concentrated on the development of effective drug-protein carriers. Much of the experimental effort has been devoted to the development of suitable drug-carrier linkages, and amide bonds between the amino group of daunomycin and the carboxylic side chains of succinylated serum albumin have proved useful for this purpose [181]. When daunomycin is linked directly to serum albumin the resulting derivatives are not biologically active. Therefore, peptidyl derivatives of daunomycin [182] are used as precursors for binding. Serum albumin-Ala-Leu-Ala-Leu-daunomycin is readily hydrolysed by lysosomal enzymes in vitro but resistant to hydrolysis in serum. The in vitro release of daunomycin also correlates well with the ability of conjugates to treat experimental L1210 leukemia in mice [76].

The problem common to all drug delivery systems is the ability to target them to specific cell types. Antibodies to specific membrane antigens may provide a solution as it is known that certain tumour cells express embryonic antigens on their surfaces which are absent from normal cells. Efforts are being made to raise and characterise tumour-specific antibodies using the modern techniques available for the production of monoclonal antibodies. There is considerable evidence that drug-antibody conjugates are therapeutically effective [183] but the mechanism of action is still under some debate. Antitumor antibodies themselves can be effective in the supression of tumour development, and it has been shown that even when administered separately, tumour-specific antibodies and drugs can act synergistically [184]. As mentioned earlier, the main difficulty associated with the use of drug-antibody conjugates is their fixation without concomitant loss in antibody/antigen affinity at the cell surface or irreversible drug inactivation. Thus, "polymeric bridges" (e.g. poly(glutamic acid) and dextran) have been used as intermediates between antibody and drug to overcome these problems.

Poznansky et al. have recently reported interesting work describing the use of albumin polymers as carriers of enzymes (for enzyme replacement therapy or as antitumour agents [185-187]). Enzyme-albumin polymers have been prepared using glutaraldehyde as a crosslinking agent and it has been shown that α-1,4-glucosidase-polymer containing antibodies raised against isolated rat hepatocytes can be effectively targeted to these cells after intravenous injection. This may be a useful approach for the treatment of the lysosomal storage disease in which the liver lysosomes become engorged with glycogen (Pompe's disease) [186]. In addition to the potential for targeting, the enzyme-albumin polymer has increased resistance to heat denaturation (compared with free enzymes) and proteolysis by trypsin. L-Asparaginase-albumin polymer is also more resistant to bioinactivation and, when administered to experimental tumours, more effective than free drug in inhibiting cell growth in vitro and prolonging survival times in vivo [187].

Now let us briefly consider the role of targetable synthetic polymers in light of all the other potential systems that are available. What advantages, if any, does the synthetic polymer provide? Let us first summarise the main advantages and disadvantages of the competitors. Liposomes represent a variety of different, relatively non-toxic, biodegradable carriers which can be used to encapsulate a large number of compounds either in the lipid or aqueous phase. Lipid compositions can be chosen such that they are stable in body fluids and the lipid may be modified in such a way as to produce altered body distributions after in vivo administration. The main drawbacks relate to liposome size and the difficulties in controlling their stability with resultant loss of control of location and rate of drug release. Natural macromolecules which have an appropriate but discreet size can be chosen. Since they are native to the physiological environment in which they will be used they are biodegradable and generally non-toxic. However, they are often immunogenic. They can also be selected such that their normal molecular structure incorporates required targeting properties e.g. antibodies, glycoproteins. Frequently, such macromolecules are extremely sensitive to denaturation and difficulties can arise in drug binding. Direct drug binding has proved particularly unsuccessful and when a series of reactions are necessary to construct a functionally active conjugate the chemistry often presents difficulties.

Although synthetic macromolecules are themselves foreign materials in the body, it is possible to select polymers which are non-toxic and non-immunogenic. Materials can be synthesized which are either resistant to hydrolysis or biodegradable, depending on the specific requirements of the respective carrier. Synthetic polymer chemistry is now sufficiently versatile to permit the development of polymeric carriers which are tailor-made, thus including the most appropriate polymer-drug linkages and targeting residues. In contrast to vesicular carriers, which intrinsically have a minimum feasible size (below which the surface tension in the shell will not allow formation of a sphere), synthetic polymers can be prepared to give any required carrier size. Of course, polymeric preparations are always polydispese, but techniques are available which allow a minimum polydispersity if this is required.

4.2 Medical Applications

Due to the difficulties in developing specific pharmacological agents many diseases would be more effectively treated if drugs could be directed with the aid of carriers to their specific site of action. Cancer chemotherapy is an obvious example where drug-targeting would alleviate many of the unpleasant side effects resulting from this treatment and many of the reported investigations are concerned with the development of polymeric systems which will hopefully improve cancer chemotherapy in the clinic in the near future.

Carrier-mediated drug delivery is not only beneficial to this particular group of diseases. Many inborn errors of metabolism, particularly lysosomal storage diseases, may be attributed to the absence of one or more key enzymes. This results in a build up of the natural substrate for the "missing" enzyme leading to pathological states of varying severity. Lysosomotropic drug carriers would therefore be ideally suitable for combating these storage diseases.

Certain parasitic diseases may also prove suitable targets for carrier-directed therapy and there are already reports on the use of drug carriers for delivering agents for the treatment of certain types of malaria [188], leishmaniasis [189] and trypanosomiasis [190]. The list of potential disorders amenable to directed chemotherapy could go on ad infinitum, but there is no doubt that with increasing costs in the development and marketing of novel pharmacological agents there will be a growing interest in the improvement of the therapeutic effectiveness of existing compounds.

5 Conclusions

Although conceptually appreciated for a number of years until now the development of soluble polymers as functionally effective carriers of therapeutic agents has moved forward relatively slowly. With the realization that polymers must be rationally designed if they are to fulfil specific biological functions, this area of research should progress much more rapidly. The versatility of polymer chemistry permits the synthesis of tailor-made macromolecules which offer certain advantages over other drug delivery systems. However, in the long run each type of carrier vehicle will no doubt find application in fields where it appears most beneficial. Coordination of the knowledge available in all areas of polymer science, biology, medicine, and pharmacology is essential for rapid progress in this field.

6 List of Abbreviations

DIVEMA   Divinyl ether and maleic anhydride copolymer
DMSO   Dimethyl sulfoxide
DOPA   3,4-Dihydroxyphenylalanine

DSC   Differential scanning calorimetry
HMDA   Hexamethylenediamine
IME   2-Imino-2-methoxyethyl-1-thioglycoside
MTX   Methotrexate
NAp   p-Nitroaniline
PAH   Poly(acryl hydrazide)
PDM   p-Phenylenediamine mustard
PEG   Poly(ethylene glycol)
PEO   Poly(ethylene oxide)
PGA   Poly(glutamic acid)
poly(Lys)   Polylysine
poly(vA)   Polyvinyladenine
poly(vU)   Polyvinyluracil
PVP   Polyvinylpyrrolidone

7 References

1. Donaruma, L. G., Vogl, O.: Polymeric Drugs, London, Academic Press Inc. 1978
2. Baker, R.: Controlled Release of Bioactive Materials, New York Academic Press Inc. 1980
3. Lewis, D. H.: Controlled Release of Pesticides and Pharmaceuticals, New York, Plenum Press 1981
4. Chang, T. M. S.: Biomedical Applications of Immobilized Enzymes and Proteins, Vols. 1 and 2, New York, Plenum Press 1977
5. Goldberg, E. P., Nakajima, A.: Biomedical Polymers, New York, Academic Press 1980
6. Gebelein, C. G., Koblitz, F. F.: Biomedical and Dental Applications of Polymers, Polymer Science and Technology Vol. 14, New York, Plenum Press 1981
7. Chiellini, E., Guisti, P.: Polymers in Medicine: Biomedical and Pharmacological Applications, Plenum Press 1983, in press
8. Heller, J., Baker, R. W.: Theory and practice of controlled drug delivery from bioerodible polymers, in: Controlled Release of Bioactive Materials (ed.) Baker, R. W., p. 1, New York, Academic Press 1980
9. Ehrlich, P.: Studies in Immunity, New York, Wiley 1906
10. Ringsdorf, H.: J. Polym. Sci. Polym. Symp. *51*, 135 (1975)
11. Pitha, J., Kusiak, J. W.,: Biological activities and targeting of soluble macromolecules, in: Controlled Release of Pesticides and Pharmaceuticals (ed.) Lewis, D. H., p. 67, New York, Plenum Press 1981
12. Allison, A. C., Davies, P.: Mechanisms of endocytosis and exocytosis, in: Transport at the Cellular Level, Society for Exp. Biol. Symp. *xxviii*, p. 419, Cambridge, Cambridge University Press 1974
13. Silverstein, S. C., Steinman, R. M., Cohn, Z. A.: Ann. Rev. Biochem. *46*, 669 (1977)
14. Stossel, T. P.: Endocytosis, in: Receptors and Recognition, Series A, (eds.) Cuatrecasas, P., Greaves, M. F. Vol 4, p. 105, London, Chapman and Hall 1977
15. Griffin, F. M. et al.: J. Exp. Med. *142*, 1263 (1975)
16. Lewis, W. H.: John Hopkins Hosp. Bull. *49*, 17 (1931)
17. Schneider, Y.-J. et al.: J. Cell Biol. *82*, 449 (1979)
18. Barrett, A. J., Heath, M. F.: Lysosomal enzymes, in: Lysosomes, a laboratory handbook (ed.) Dingle, J. T., p. 19, Amsterdam, Elsevier/North-Holland Biomedical Press 1977
19. Jacques, P.: Endocytosis, in: Lysosomes in Biology and Pathology, Vol. 2 (eds.) Dingle, J. T., Fell, H. B., p. 395, Amsterdam, North Holland Biomedical Press 1969
20. Neufeld, E. F., Ashwell, G.: Carbohydrate recognition systems for receptor-mediated pino-

cytosis, in: The Biochemistry of Glycoproteins and Proteoglycans (ed.) Lennarz, W. S., p. 241, New York, Plenum Press 1980
21. Goldstein, J. L., Brown, M. S.: Ann. Rev. Biochem. *46*, 897 (1977)
22. De Duve, C. et al.: Biochem. Pharmacol. *23*, 2495 (1974)
23. Reijngoud, D.-J., Tager, J. M.: Biochim. Biophys. Acta *472*, 419 (1977)
24. Lloyd, J. B.: Experimental support for the concept of lysosomal storage diseases, in: Lysosomes and Storage Diseases (eds.) Hers, H. G., Van Hoof, F., p. 173, New York and London, Academic Press 1973
25. Docherty, K. et al.: Biochem. J. *178*, 361 (1979)
26. Kopeček, J.: Soluble polymers in medicine, in: Systemic Aspects of Biocompatibility (ed.) Williams, D. F., p. 159, Boca Raton, Florida, CRC Press 1981
27. Zaharko, D. S., Przybylski, M., Oliverio, V. T.: Meth. Canc. Res. *16*, 347 (1979)
28. Šprincl, L. et al.: J. Biomed. Mater. Res. *10*, 953 (1976)
29. Duncan, R. et al.: Biochem. J. *196*, 49 (1981)
30. Cartlidge, S. A. et al.: in preparation
31. Cowie, J. M. G.: Polymers: Chemistry and Physics of Modern Materials, Aylesbury, UK, Int. Textbook Co. 1973
32. Kopeček, J., Ulbrich, K.: Progr. Polym. Sci., *9*, 1 (1983)
33. Kopeček, J., Rejmanová, P.: Enzymatically degradable bonds in synthetic polymers, in: Controlled Drug Delivery (ed.) Bruck, S. D., p. 81, Boca Raton, Florida, CRC Press 1983
34. Kopeček, J.: Makromol. Chem. *178*, 2169 (1977)
35. Kopeček, J.: Biodegradation of polymers for biomedical use, in: IUPAC Macromolecules (ed.) Benoit, H., Rempp, P., p. 305, Oxford, Pergamon Press 1982
36. Rejmanová, P., Obereigner, B., Kopeček, J.: Makromol. Chem. *182*, 1899 (1981)
37. Sekiguchi, H.: Pure Appl. Chem. *53*, 1689 (1981)
38. Vasiliev, A. E.: Medical polymers (in Russian), in: Itogi nauki i techniki *16*, 3 (1981)
39. Gros, L., Ringsdorf, H., Schupp, H.: Angew. Chem. *93*, 311 (1981)
40. Morawetz, H.: J. Polym. Sci., Polym. Symp. *62*, 271 (1978)
41. Mikeš, F. et al.: Macromolecules *14*, 175 (1981)
42. Kaplan, A. M.: Antitumor activity of synthetic polyanions, in: Anionic Polymeric Drugs (ed.) Donaruma, L. G., Ottenbrite, R. M., Vogl, O., p. 227, New York, J. Wiley 1980
43. Ottenbrite, R. M. et al.: Biological activity of poly (carboxylic acid) polymers, in: Polymeric Drugs (ed.) Donaruma. L. G., Vogl. O., p. 263, New York, Academic Press 1978
44. Hespe, W., Meier, A. M., Blankwater, Y. J.: Drug Res. *27*, 1158 (1977)
45. Butler, G. B.: Synthesis, characterization, and biological activity of pyran copolymers, in: Anionic Polymeric Drugs (ed.) Donaruma, L. G., Ottenbrite, R. M., Vogl, O., p. 49, New York, J. Wiley 1980
46. Mück, K. F., Rolly, H., Burg, K.: Makromol. Chem. *178*, 2773 (1977)
47. Mück, K. F., Christ, O., Kellner, H. M.: Makromol. Chem. *178*, 2785 (1977)
48. Allcock, H. R. et al.: Macromolecules *10*, 824 (1977)
49. Grolleman, C. W. J. et al.: Polyphosphazenes as a system for programmed drug release, in: Proceedings of the Int. Conf. Biomedical Polymers, p. 203, London, The Biological Engineering Society 1982
50. Kopeček, J., Šprincl, L., Lím, D.: J. Biomed. Mater. Res. *7*, 179 (1973)
51. Hofmann, V., Ringsdorf, H., Muacevic, G.: Makromol. Chem. *176*, 1929 (1975)
52. Carpino, L. A., Ringsdorf, H., Ritter, H.: Makromol. Chem. *177*, 1631 (1976)
53. Neri, P. et al.: J. Med. Chem. *16*, 893 (1973)
54. Antoni, G. et al.: Biopolymers *13*, 1721 (1974)
55. Petersen, R. V. et al.: Biodegradable drug delivery systems based upon poly(L-glutamic acid) and poly(L-glutamines), in: Proceedings of the Int. Conf. Biomedical Polymers, p. 211, London, The Biological Engineering Society 1982
56. Pitha, J.: Polymeric drugs: Effects of polyvinyl analogs of nucleic acids on cells, animals and their viral infections, in: Biomedical and Dental Applications of Polymers (ed.) Gebelein, C. G. Koblitz, F. K., p. 203, New York, Plenum Press 1981
57. Takemoto, K.: Recent problems concerning functional monomers and polymers containing nucleic acid bases, in: Polymeric Drugs (ed.) Donaruma, L. G., Vogl, O., p. 103, London, Academic Press 1978

58. Ajisaka, K., Iwashita, Y.: Biochem. Biophys. Res. Commun. *97*, 1076 (1980)
59. Franzmann, G., Ringsdorf, H.: Makromol. Chem. *177*, 2547 (1976)
60. Ferrutti, P.: Il Farmaco, Ed. Sci. *32*, 220 (1977)
61. Hörpel, G. et al.: Micellforming co- and blockcopolymers for sustained drug release, in: Proceedings of the 28th IUPAC Macromolecular Symposium, p. 346, Amherst, Ma. 1982
62. Duncan, R., et a.: Biochim. Biophys. Acta *717*, 248 (1982)
63. Duncan, R., et al.: unpublished results
64. Hofmann, V., Ringsdorf, H., Schaumlöffel, E.: Makromol. Chem. *181*, 351 (1980)
65. Fuller, W. D., Verlander, M. S., Goodman, M.: Biopolymers *17*, 2939 (1978)
66. Pitha, J. et al.: Makromol. Chem. *182*, 1945 (1981)
67. Carraher, C. E.: Organometallic polymers as drugs and drug delivery systems, in: Biomedical and Dental Applications of Polymers (ed.) Gebelein, C. G., Koblitz, F. K., p. 215, New York, Plenum Press 1981
68. Carraher, C. E. et al.: J. Macromol. Sci. Chem. *A15*, 625 (1981)
69. Carraher, C. E., Moon, W. G., Langworthy, T. A.: Polymer Preprints p. 1, ACS Spring Meeting 1976
70. Molz, P.: Synthese und Untersuchung von potentiell spaltbaren Spacergruppen zur Polymerfixierung von NOR-Stickstoff-LOST und den Anthracyclinen Daunomycin und Adriamycin, Ph. D. Thesis, Johannes Guttenberg University Mainz, FRG 1982
71. Morawetz, H.: Macromolecules in Solution, New York, Interscience Publishers 1975
72. Shen, W. C., Ryser, H. J. P.: Biochem. Biophys. Res. Commun. *102*, 1048 (1981)
73. Kopeček, J., Rejmanová, P., Chytrý, V.: Makromol. Chem. *182*, 799 (1981)
74. Duncan, R. et al.: Bioscience Reports, *2*, 1041 (1982)
75. Rejmanová et al.: Makromol. Chem., *184*, 000 (1983) No. 10 — October
76. Baurain, R. et al.: Antitumoral activity of daunorubicin linked to proteins. II. Lysosomal hydrolysis and antitumoral activity of conjugates prepared with peptidic spacer arms, in: Proceecings of the 12th Congress of Chemotherapy, Florence 1981
77. Duncan, R., Kopeček, J., Lloyd, J. B.: Development of N-(2-hydroxypropyl)methacrylamide copolymers as carriers of therapeutic agents, in: Polymers in Medicine: Biomedical and Pharmacological Applications (eds.) Chiellini, E., Giusti, P., New York, Plenum Press 1983
78. Duncan, R. et al.: Biochim. Biophys. Acta *678*, 143 (1981)
79. Rejmanová, P. et al.: unpublished results
80. Molz, P. et al.: Int. J. Biol. Macromol. *2*, 245 (1980)
81. Hofmann, V. et al.: Makromol. Chem. *180*, 837 (1979)
82. Obereigner, B. et al.: J. Polym. Sci. Polym. Symp. *66*, 41 (1979)
83. Sheehan, J. C., Hess, G. P.: J. Am. Chem. Soc. *77*, 1067 (1955)
84. Wieland, T., Kern, W., Sehring, R.: Ann. Chem. *569*, 117 (1950)
85. Przybylski, M. et al.: Makromol. Chem. *179*, 1719 (1978)
86. Rejmanová, P., Labský, J., Kopeček, J.: Makromol. Chem. *178*, 2159 (1977)
87. Říhová, B. et al.: Biomaterials in press *184*, 1345 (1983)
88. Pitha, J., Zawadski, S., Hughes, B. A.: Makromol. Chem. *183*, 781 (1982)
89. Chytrý, V., Kopeček, J.: Makromol. Chem., *184*, 1345 (1983)
90. Hirano, T., Klesse, W., Ringsdorf, H.: Makromol. Chem. *180*, 1125 (1979)
91. Lääne, A. et al.: Collect. Czech. Chem. Commun. *46*, 1466 (1981)
92. Joost, H. G., Hasselblatt, A.: Naunyn-Schmiedeberg's Arch. Pharmacol. *297*, 81 (1977)
93. Trouet, A.: Europ. J. Cancer *14*, 105 (1978)
94. Lee, Y. C., Stowell, C. P., Krantz, M. J.: Biochemistry *15*, 3956 (1976)
95. Lee, R. T., Lee, Y. C.: Biochemistry *19*, 156 (1980)
96. Kawaguchi, K. et al.: J. Biol. Chem. *256*, 2230 (1981)
97. Kawaguchi, K. et al.: Arch. Biochem. Biophys. *205*, 388 (1980)
98. Duncan, R. et al.: Biochim. Biophys. Acta *755*, 518 (1983)
99. Hurwitz E. et al.: Cancer Res. *35*, 1175 (1975)
100. Davies, D. A. L., O'Neill, G. J.: Proc. XI. Int. Cancer Cong. *1*, 218 (1974)
101. Rowland, G. F.: Eur. J. Cancer *13*, 593 (1977)
102. Wilchek, M.: Makromol. Chem. Suppl. *2*, 207 (1979)

103. O'Neill, G. J.: The use of antibodies as drug carriers, in: Drug Carriers in Biology and Medicine (ed.) Gregoriadis, G., p. 23, London, Academic Press 1979
104. Říhová, B., Kopeček, J.: unpublished results
105. Winter, G. D., Leray, J. L., De Grost, K.: Evaluation of Biomaterials, Chichester—New York—Brisbane—Toronto, John Wiley & Sons 1980
106. Munton, C. G. F. et al.: Br. J. Ophtal. 58, 941 (1974)
107. Vert, M. et al.: Makromol. Chem., Suppl. 5, 30 (1981)
108. Duncan, R., Lloyd, J. B., Kopeček, J.: Biochem. Biophys. Res. Commun. 94, 284 (1980)
109. Duncan, R. et al.: Cell Biol. Int. Reports 5 (Suppl. A.), 14 (1981)
110. Hirano, T., Ringsdorf, H., Zaharko, D. S.: Cancer Res. 40, 2263 (1980)
111. Drobník et al. Makromol. Chem. 177, 2833 (1976)
112. Fu, T.-Y., Morawetz, H.: J. Biol. Chem. 251, 2083 (1976)
113. Drobník, J. et al.: J. Polym. Sci. Polym. Symp. 66, 65 (1979)
114. Verlander, M. et al.: Some novel approaches to the design and synthesis of peptide-catecholamine conjugates in: Polymers in Medicine: Biomedical and Pharmacological Applications (eds.) Chiellini, E., Guisti, P., Plenum Press, 1983 in press
115. Ryser, H. J.-P., Shen, W.-C.: Proc. Natl. Acad. Sci. USA 75, 3867 (1978)
116. Shen, W.-C., Ryser, H. J.-P.: Mol. Pharmacol. 16, 614 (1979)
117. Shen, W.-C., Ryser, H. J.-P.: Fed. Proc. 40, 642 (1981)
118. Chu, B. C. F., Howell, S. B.: J. of Pharm. Exp. Thera. 219, 389 (1981)
119. Van Heeswijk, W. A. R. et al.: Synthesis and characterisation of a macromolecular prodrug of the antitumor antibiotic adriamycin, in: Proceedings of International Symposium on Polymers in Medicine, Porto Cervo, Sardinia, p. 23, 1982
120. Ferruti, P. et al.: Polymeric derivatives of daunorubicin as drug delivery systems in antitumor chemotherapy, in: Proceedings of the International Symposium on Polymers in Medicine, Port Cervo, Sardinia 1982
121. Williams, K. E. et al.: J. Cell Biol. 64, 123 (1975)
122. Pratten, M. K., Williams, K. E., Lloyd, J. B.: Biochem. J. 168, 365 (1977)
123. Leake, D. S., Bowyer, D. E.: Biochem. Soc. Trans. 5, 130 (1977)
124. Bridges, J. F., Woodley, J. F. in: Maternofoetal transmission, Vol. 2, (ed.) Hemming, W., p. 249, Amsterdam, Elsevier 1979
125. Rowland, R. N., Woodley, J. F.: Bioscience Reps. 1, 399 (1981)
126. Breslow, D. S.: Pure Appl. Chem. 46, 103 (1976)
127. Pratten, M. K. et al.: Chem.-Biol. Interactions 35, 319 (1981)
128. Papamatheakis et al.: Cancer Treat. Rep. 62, 1845 (1978)
129. Seljelid, R., Silverstein, S. C., Cohn, Z. A.: J. Cell Biol.57, 484 (1973)
130. Ryser, H. J.-P.: Nature 215, 934 (1967)
131. Ryser, H. J.-P., Shen, W.-C., Merk, F. B.: Life Sciences 22, 1253 (1978)
132. Pratten, M. K., unpublished results
133. Chu, B. C. F., Howell, S. B.: Biochem. Pharmacol. 30, 2545 (1981)
134. Noronha-Blob, L. et al.: J. Med. Chem. 20, 356 (1977)
135. Tirrell, D. A., Boyd, P. M.: Makromol. Chem. Rapid Commun. 2, 193 (1981)
136. Pratten, M. K. et al.: Biochim. Biophys. Acta 719, 424 (1982)
137. Cartlidge, S. A. et al., in: Proceedings of the Int. Conf. Biomedical Polymers, p. 289, London, The Biological Engineering Society 1982
138. Fornůsek, L., Větvička, V., Kopeček, J.: Experientia 37, 418 (1981)
139. Ravin, H. A., Seligman, A. M., Fine, J.: New England J. Med. 247, 921 (1952)
140. Regoeczi, E.: Br. J. exp. Path. 57, 431 (1976)
141. Munniksma, J. et al.: Biochem. J. 192, 613 (1980)
142. Ashwell, G., Morell, A. G.: Adv. Enzymol. 41, 99 (1974)
143. Stahl, P. D. et al.: Proc. Natl. Acad. Sci. 75, 1399 (1978)
144. Achord, D. T. et al.: Cell 15, 269 (1978)
145. Kaplan, A. et al.: J. Clin. Invest. 60, 1088 (1977)
146. Abel, G. et al.: Makromol. Chem. 177, 2669 (1976)
147. Ohno, H., Abe, K., Tsuchida, E.: Makromol. Chem. 182, 1253 (1981)
148. Lenz, R. W., Guerin, P.: Functional polyesters and polyamides for medical applications of

biodegradable polymers, in: Polymers in Medicine: Biomedical and Pharmacologial Applications (eds.) Chiellini, E., Guisti, P. Plenum Press 1983 in press

149. Braswell et al.: The synthesis and titration of poly(amino acids), in: Proceedings of the International Symposium on Polymers in Medicine, Porto Cervo, Sardinia, p. 36, 1982
150. Kopeček, J. et al.: Makromol. Chem. *182*, 2941 (1981)
151. Fauvarque, J. F., Malinge, J.: Synthesis of biodegradable hydrosoluble polymers, in: Proceedings of the International Symposium on Porto Cervo, Sardinia p. 41, 1982
152. Williams, D. F.: Introduction to the toxicology of polymer-based materials, in: Systemic Aspects of Biocompatibility (ed.) Williams, D. F. *Vol. II*, 51, Boca Raton, Florida, CRC Press 1981
153. Abuchowski, A. et al.: J. Biol. Chem. *252*, 3578 (1977)
154. Savoca, K. V. et al.: Biochim. Biophys. Acta *578*, 47 (1979)
155. Goldman, I. D.: Cancer Chemother. Rep. *6*, 63 (1975)
156. Chu, B. C. F., Howell, S. B.: J. Pharm. Exp. Thera. *219*, 389 (1981)
157. Fung et al.: J. Natl. Cancer Inst. *62*, 1261 (1979)
158. Przybylski, M. et al.: Cancer Treatment Reps. *62*, 1837 (1978)
159. Hörpel, G. et al.: Micelle-forming copolymers and block copolymers for sustained drug release, in: Proceeding of the International Symposium on Polymers in Medicine, Porto Cervo, Sardinia 1982
160. Hurwitz, E., Wilchek, M., Pitha, J.: J. Appl. Biochem. *2*, 25 (1980)
161. Von Sprecht, B.-U., Seinfeld, H., Brendel, W.: Hoppe-Seyler's Z. Physiol. Chem. *354*, 1659 (1973)
162. Hespe, W., Blankwater, Y. J., Wieriks, J.: Arzneim.-Forsch. *25*, 1561 (1975)
163. Gregoriadis, G.: Drug Carriers in Biology and Medicine, London, Academic Press Inc. 1979
164. Knight, C. G.: Liposomes: From Physical Structure to Therapeutic Applications, Amsterdam, North-Holland Biomedical Press 1981
165. Nicholls, P.: Liposomes — as artifical organelles, topochemical matrices, and therapeutic carrier systems, in: Membrane Research, Classic Origins and Current Concepts (ed.) Muggleton Harris, A. L. p. 327, New York, Academic Press Inc. 1981
166. Kirby, C., Clarke, J., Gregoriadis, G.: Biochem. J. *186*, 591 (1980)
167. Sunamoto, J. et al.: Improved drug delivery to target-specific organs using liposomes as coated with polysaccharide, in: Proceedings of the International Symposium on Polymers in Medicine, Porto Cervo, Sardinia, p. 3, 1982
168. Couvreur, P. et al.: Febs. Lett. *84*, 323 (1977)
169. Poste, G., Fidler, I. J.: Therapeutic amplification of macrophage-mediated destruction of tumor cells, an approach to cancer chemotherapy that addresses the problem of tumor cell heterogeneity, in: Design of Models for Testing Cancer Therapeutic Agents (eds.) Fidler, I. J., White, R. J., p. 225, New York, Van Nostrand Reinhold 1982
170. Gregoriadis, G.: N. Engl. J. Med. *295*, 704 (1976)
171. Gregoriadis, G. et al.: Lancet *1*, 1313 (1974)
172. Tyrrell, D. A. et al.: Brit. Med. J. *2*, 88 (1976)
173. Weismann, G. et al.: Ann. N. Y. Acad. Sci. *308*, 235 (1978)
174. Mantovani, P., Pepeu, G., Amaducci, L.: Adv. Exp. Med. Biol. *72*, 285 (1976)
175. Masturzo, P. et al.: New Engl. J. Med. *297*, 338 (1977)
176. Zwall, R. F. A.: Biochim. Biophys. Acta *515*, 163 (1978)
177. Allison, A. C., Gregoriadis, G.: Nature *252*, 252 (1974)
178. Trouet, A. et al.: DNA, liposomes and proteins as carriers for antitumoral drugs, in: Recent Results in Cancer Research, Vol. *75* (eds.) Mathe, G., Muggia, F. M. p. 229, Berlin—Heidelberg, Springer-Verlag 1980
179. Trouet, A., Deprez-De Campeneere, D., De Duve, C.: Nature New Biol. *239*, 110 (1972)
180. Trouet, A., Sokal, G.: Cancer Chemother. Rep. *63*, 895 (1979)
181. Trouet, A. et al.: Proc. Natl. Acad. Sci. USA *79*, 626 (1982)
182. Masquelier, M. et al.: Antitumoral activity of daunorubicin linked to proteins. I. Biological and antitumoral properties of peptidic derivatives of danorubicin used as intermediates, in: Proceedings of the 12th Congress of Chemotherapy, Florence 1981
183. Möller, G.: Antibody Carriers of Drugs and Toxins in Tumor Therapy Immunol. Rev. *62*, Copenhagen, Munksgaard 1982
184. Newman, C. E. et al.: Lancet *1*, 163 (1977)

185. Poznansky, M. J., Bhardwaj, D.: Can. J. Physiol. Pharmacol. *58*, 322 (1980)
186. Poznansky, M. J., Bhardwaj, D.: Biochem. J. *196*, 89 (1981)
187. Poznansky, M. et al.: Cancer Res. *42*, 1020 (1982)
188. Alving, C. R. et al.: Science *205*, 1142 (1979)
189. New, R. R. C. et al.: Nature *272*, 55 (1978)
190. Gruenberg, J. et al.: Biochem. Biophys. Res. Commun. *88*, 1173 (1979)

Karel Dušek (Editor)
Received June 15, 1983

Blood-Compatible Polymers

Yoshito Ikada
Research Center for Medical Polymers and Biomaterials,
Kyoto University, Kyoto 606, Japan

This article describes polymers which are blood-compatible not by the addition of anticoagulants or due to the formation of neointima, but because they consist of antithrombogenic biomaterial that does not include blood-soluble additives either. Since research in this area is still in its infancy, the author has placed the emphasis on his own experimental results.

Advances in Polymer Science 57
© Springer-Verlag Berlin Heidelberg 1984

1 Abbreviations

Monomer

AAm	acrylamide
HEMA	2-hydroxyethyl methacrylate
MMA	methyl methacrylate
N-VP	N-vinyl pyrrolidone
DMAEM	2-(N, N-dimethylamino)ethyl methacrylate
AA	acrylic acid
AANa	sodium acrylate

Polymer

PAAm	polyacrylamide
PVA	poly(vinyl alcohol)
PAA	poly(acrylic acid)
PE	polyethylene
HDPE	high-density polyethylene
LDPE	low-density polyethylene
PP	polypropylene
PHEMA	poly(2-hydroxyethyl methacrylate)
PMMA	poly(methyl methacrylate)
PET	poly(ethylene terephthalate)
PTFE	polytetrafluoroethylene
VAECO	vinyl alcohol-ethylene copolymer
BSA	bovine serum albumin

Miscellaneous

PRP	platelet-rich plasma
AIVC	4,4'-azobis-4-cyanovaloyl chloride
TMPO	2,2,6,6-tetramethyl-4-aminopiperidine-1-oxyl
ESR	electron spin resonance
XPS	X-ray photoelectron spectroscopy
ESCA	electron spectroscopy for chemical analysis
SEM	scanning electron microscopy

2 Introduction

So far, it has been widely believed that any surface of synthetic polymers sooner or later causes blood clotting when it comes into contact with the fresh blood. However, polymers are increasingly used in medicine even in cases in which the polymer surface is in direct contact with blood. Typical examples are blood bags and membranes in extracorporeal blood circulation. In these cases the period of their contact with blood is relatively short and hence thrombus formation can be avoided by the addition of Ca^{2+}-binders or biologically active polymers, such as heparin, to blood. The use of these anticoagulants, however, is not allowed when polymeric materials are implanted for a longer period, as in vascular grafts. In such cases the material should be made blood-compatible without any releasable, biologically active polymers.

This may be achieved by neointima formation or the use of antithrombogenic biomaterial. In the former case the thrombi formed on the surface are expected to organize into the pseudoneointima, which should be fixed on the material surface and viable for a long period. This is the common method used for vascular grafts of arteries with a diameter larger than about 6 mm. For substitutes of veins and arteries with a smaller diameter, this stable neointima formation is hardly expected because the conduit undergoes occlusion by the thrombi formed almost instantly when the surface comes into contact with blood, or because the neointima grows continuously to occlude entirely the conduit of small diameter.

However, there are some cases where such neointima formation is not desirable. For instance, the biosensor of artificial pancreas and the bioadsorbent of artificial liver would undergo substantial reduction in efficiency if blood components deposit on the surface.

This article describes polymer surfaces the blood compatibility of which is achieved neither by the use of heparin nor by the formation of neointima, but by the antithrombogenic surface by itself. Thus, a blood-compatible polymer here means an *antithrombogenic biomaterial* and not a water-soluble polymer which is completely soluble in blood and works as plasma expander or anticoagulant.

3 Overview of Current Blood-Compatible Polymers

3.1 Classification of Blood-Compatible Polymers

Segmented *polyetherurethanes* contain no water-soluble additives and are characterized by a high hydrolytic resistance, good processability, and good mechanical properties in comparison with other elastomers. Therefore, polyurethanes of this type are widely used in medicine, for example as aortic balloon pumps, catheters, and housings for artificial hearts [1]. Thus, a correlation between blood compatibility and the chemical structure of polyurethanes has often been reported [2, 3].

Silicone has also a long history as a biomedical polymer and has preferably been used for special catheters and shunts which require good blood compatibility. This is because the polymer exhibits an extremely high chemical stability and hydrophobicity. Poor mechanical properties can be improved by mixing with fillers, which, however, decrease the blood compatibility [4]. A commercial product, Avcothane®, seems to be composed largely of polyurethane and silicone [5].

Polytetrafluoroethylene (PTFE) is chemically more inert and displays a higher hydrophobicity than silicone. *Expanded PTFE* with its microporous structure is often used for vascular grafts and patches. Normally, vascular grafts made from PTFE become blood-compatible as a result of neointima ingrowth, but are also reported to exhibit relatively good antithrombogenicity if the micropores are filled with plenty of water [6].

In contrast to the above mentioned hydrophobic polymers, the so-called *hydrogel*, which is a physically or chemically crosslinked polymer swollen with water, is hydrophilic and also known as a blood-compatible polymer [7]. *Poly(2-hydroxyethyl meth-*

acrylate) (PHEMA) is a typical hydrogel polymer with a water content of about 50 wt-% [8]. However, this polymer exhibits poor mechanical properties. Therefore, it is mostly used for coating or grafting onto a mechanically strong substrate.

A pyrolytic, *low-temperature isotropic* (*LTI*) *carbon* is also used by depositing it on substrates such as metals. For instance, an artificial heart valve is composed of a metal coated with LTI carbon. It has not been found out so far why this carbon has a good blood compatibility [9].

Natural tissues such as bovine carotid artery and human umbilical cord vein have been clinically used as thromboresistant vascular prosthesis after treatment with aldehydes like dialdehyde starch and glutaraldehyde. Coating with gelatin followed by glutaraldehyde crosslinking is also reported to give a nonthrombogenic surface, called *biolized surface*, and used on lining of blood pumps [10]. Precoating with serum albumin provides a good method for creating nonthrombogenicity, though not suitable for long-term use [11].

3.2 Factors Influencing Blood Compatibility of Polymers

Numerous controversial theories on blood-compatible polymers have been reported, because the interactions of blood with polymer surfaces leading to thrombus formation are governed by many chemical, physical, and biological parameters [12] and difficult to analyse in detail. The important variables affecting thrombus formation on biomaterials include: chemical composition of the surface, physical texture of the surface, disturbance of blood flow induced by the polymer, the kind of animals used, the physical conditions and local place of the tested body, etc. Furthermore, the *evaluation methods* employed for blood compatibility are important because different evaluation methods lead to different conclusions; therefore, a standard sample is strongly desired [13]. At least, in the final test stage it is essential to evaluate the blood compatibility under conditions applied in practice.

The primary approach widely adopted for investigating the blood compatibility of polymers is to vary the molecular properties of the polymer surface while keeping other parameters constant. This means that the surface structure of the polymer is regarded as a dominant factor of blood compatibility.

3.3 Recent Hypotheses about Blood-Compatible Surfaces

The polymer surface influencing the blood compatibility should be discussed in terms both of the *physical* and *chemical structure*. The physical structure is mainly concerned with roughness and porosity of the surface texture. It is generally accepted that the smoother the polymer surface, the more antithrombogenic it is [14]. Physical defects on the surface, such as small pinholes and grooves, may trap microemboli and perturb the laminar flow of blood to result in thrombus formation.

The chemical structure of smooth polymer surfaces is specified by ionogenicity, hydrophobicity, hydrophilicity, and their distribution. Polymer blood compatibility is undoubtedly dependent on the chemical structure. Therefore, a variety of hypotheses about the optimum blood-compatible surface exist.

One group of investigators considers the surface free energy as very important. Bair [15] has concluded that the polymer surface will be blood-compatible if it has a critical surface tension ranging between 20 and 25 dyne · cm^{-1}, [16] while Andrade formerly postulated that the smaller the interfacial free energy between blood and the polymer surface, the better the blood compatibility [17]. Recently, he and his coworkers have reported that the blood compatibility is not dependent on only one surface parameter [18]. Ratner and coworkers have pointed out that excellent blood compatibility demands a well-balanced hydrophilicity and hydrophobicity [19]. With respect to the chemical nature of the soft segments in polyurethane, Hanson and coworkers stated that the soft segments should be hydrophobic to exhibit good blood compatibility [20], whereas Merrill and coworkers found the reverse trend on the basis of X-ray photoelectron spectroscopy (XPS) [21].

Other groups do not put much emphasis on the surface energy but pay more attention to the ionic nature of the polymer surface. Sawyer and Srinivasan assume that an anionically charged surface has a good blood compatibility since repulsive interactions are operative between the surface and platelets which also possess anionic charge [22]. The importance of electrical properties in blood compatibility was suggested by Bruck [23]. Furthermore, there is a hypothesis emphasizing a microphase structure of the surface [24]. A well-known assumption was proposed by Kim and his coworkers saying that the polymer surface selectively adsorbing serum albumin may be nonthrombogenic [25].

The objective of this article is not to give a detailed review of the theories but to unify the above diverse hypotheses as much as possible. For this purpose, we will refer mostly to the results observed in our laboratories. General aspects of blood interaction with polymer surfaces are described in other publications [26-29].

4 Adhesion in Aqueous Media

For the molecular designing of blood-compatible polymers, it is essential to know the mechanism of clotting occurring on foreign surfaces. Due to recent progress made in biochemistry of blood coagulation, the primary reactions occurring in the complicated cascade process of thrombus formation have been elucidated [30]. Although platelet adhesion as well as the activation of the Hageman factor (factor XII) in which high-molecular-weight kininogen and prekallikrein are involved [31] are not completely clear, there is no reason to suspect that thrombus formation is triggered by interactions between blood components and the foreign polymer surface [12]. It has not been found out so far which blood component plays a decisive role in the trigger reaction, but it is highly probable that some plasma proteins participate in the key reaction since protein adsorption takes place within a few seconds after a polymer surface has been brought into contact with blood [32]. The most important event in thrombus formation on biomaterials, i. e., *platelet adhesion*, seems to follow *protein adsorption*. Therefore, the adsorption of plasma proteins on the polymer surface has been the subject of numerous investigators [33].

First of all, it should be pointed out that protein adsorption is not a specific biological event but rather a physicochemical process. In addition, among a large number of events resulting in thrombus formation, physicochemical adsorption is the only reaction that can be relatively readily controlled, unless any biologically active sub-stances are used.

Our basic assumption is that a polymer surface which does not interact with blood at all, does not induce thrombus formation. In other words, *the polymer surface, which does not adsorb any plasma protein, must be blood-compatible.*

Although there have been accumulated vast experimental results indicating that the protein adsorption depends greatly on the polymer surfaces, little is known on theoretical correlation of the surface nature with the adsorption in aqueous media. Therefore, we will attempt first to derive a theoretical expression for the work of adsorption in terms of the surface energy [34]. In general, adsorption is related to a surface interaction with a substance which can be regarded as one-dimensional point, while adhesion is a surface interaction with a substance having a two-dimensional area. Since the plasma proteins as well as the cells in blood interacting with the polymer surface are better to regard as two-dimensional, we use the term "work of adhesion" instead of work of adsorption.

In contrast to adhesion in air, only few thermodynamic studies have been devoted to adhesion in the biological system. Recently van Oss [35] has pointed out that phago-cytosis of bacteria by a phagocyte is independent of immune reactions but determined solely by the free energy change during phagocytosis:

$$\Delta F_{net} = \gamma_{PB} - \gamma_{BW} \tag{1}$$

where γ_{BW} is the interfacial energy between a bacterium and water and γ_{PB} is the inter-facial energy between the phagocyte and the bacterium. For instance, he and his coworkers [36] studied the phagocytosis of bacteria by platelets in aqueous media of different surface tensions from a surface thermodynamic aspect and showed that bacterial ingestion should increase with the increasing bacterial surface tension if the liquid medium has a surface tension lower than that of the platelets. According to Gerson and Scheer [37], the extent of adhesion of bacterial cells to hydrophobic plastic surfaces increased as the difference in the free energy accompanying the cell attach-ment became larger. The free energy change is given by

$$\Delta G_a = \gamma_{SM} + \gamma_{BM} - \gamma_{SB} \tag{2}$$

where γ_{SB}, γ_{SM}, and γ_{BM} are the interfacial free energies between the solid and the bacterium, between the solid and the aqueous medium, and between the bacterium and the aqueous medium, respectively. Such a thermodynamical approach to bacterial adhesion was also attempted by Dexter [38]. He thought that upon conditioning film adsorption, the original solid-water interface (SW) was replaced with a solid-adsorbed organic interface (SO) plus a diffuse organic-water interface (OW). Therefore, the change in the free energy is

$$\Delta F = \gamma_{SO} + \gamma_{OW} - \gamma_{SW} \tag{3}$$

In evaluating the interfacial energy between the cell and the substrate in the above equations, all of the workers do not take into consideration the aqueous environment in which the cell adhesion takes place.

4.1 Work of Adhesion in Water

It is well-known that the work of adhesion in vacuum or in air (W_{12}) is defined as

$$W_{12} = \gamma_1 + \gamma_2 - \gamma_{12} \tag{4}$$

where γ_1 and γ_2 are surface free energies of body 1 and 2, respectively, and γ_{12} is the interfacial free energy. In analogy with W_{12}, the expression for the work of adhesion in water or in aqueous media ($W_{12,w}$) may be written as

$$W_{12,w} = \gamma_{1w} + \gamma_{2w} - [\gamma_{12}]_w \tag{5}$$

where γ_{1w} (γ_{2w}) is the free energy of the interface between body 1 (body 2) and water and $[\gamma_{12}]_w$ the γ_{12} value in water. If both bodies have a surface free energy consisting of only two components, i.e., dispersive (γ^d) and polar (γ^p), then

$$\gamma_1 = \gamma_1^d + \gamma_1^p \tag{6}$$
$$\gamma_2 = \gamma_2^d + \gamma_2^p \tag{7}$$

and W_{12} is assumed to be given by [39]

$$W_{12} = 2(\gamma_1^d \gamma_2^d)^{1/2} + 2(\gamma_1^p \gamma_2^p)^{1/2} \tag{8}$$

In analogy with Eq. (8) we may obtain the following Equation for $W_{12,w}$,

$$W_{12,w} = [2(\gamma_{1w}^d \gamma_{2w}^d)^{1/2} + 2(\gamma_{1w}^p \gamma_{2w}^p)^{1/2}]_w \tag{9}$$

Here the value in square brackets denotes, as is in Eq. (5), the value in aqueous medium. It seems reasonable to express γ_{1w}^d as

$$\gamma_{1w}^d = \gamma_1^d + \gamma_w^d - 2(\gamma_1^d \gamma_w^d)^{1/2} \tag{10}$$

This expression can be rewritten as

$$\gamma_{1w}^d = \gamma_1^d \left[1 - \frac{2(\gamma_1^d \gamma_w^d)^{1/2}}{\gamma_1^d + \gamma_w^d} \right] + \gamma_w^d \left[1 - \frac{2(\gamma_1^d \gamma_w^d)^{1/2}}{\gamma_1^d + \gamma_w^d} \right] \tag{11}$$

The first term in the right-hand equation is the contribution from body 1 and the second term that from water. Assuming that equations similar to Eqs. (10) and (11) also hold for γ_{1w}^p, γ_{2w}^d, and γ_{2w}^p and inserting these equations into Eq. (9), we get

$$W_{12,w} = 2 \left\{ \gamma_1^d \left[1 - \frac{2(\gamma_1^d \gamma_w^d)^{1/2}}{\gamma_1^d + \gamma_w^d} \right] \right\}^{1/2} \left\{ \gamma_2^d \left[1 - \frac{2(\gamma_2^d \gamma_w^d)^{1/2}}{\gamma_2^d + \gamma_w^d} \right] \right\}^{1/2}$$
$$+ 2 \left\{ \gamma_1^p \left[1 - \frac{2(\gamma_1^p \gamma_w^p)^{1/2}}{\gamma_1^p + \gamma_w^p} \right] \right\}^{1/2} \left\{ \gamma_2^p \left[1 - \frac{2(\gamma_2^p \gamma_w^p)^{1/2}}{\gamma_2^p + \gamma_w^p} \right] \right\}^{1/2} \tag{12}$$

In deriving the above equation, we inserted only the first terms of Eq. (11) and the corresponding equations into Eq. (9), since the second terms of these equations are related to water and may hardly contribute to $W_{12,\,w}$.

If $[\gamma_{12}]_w$ is assumed to be equal to γ_{12}, the work of adhesion, in this case, $W_{12',w}$ is reduced to

$$W_{12',\,w} = \gamma_{1w} + \gamma_{2w} - \gamma_{12} \tag{13}$$

4.2 Comparison of Theoretical and Observed Results

Calculation of $W_{12,\,w}$ from Eq. (12) requires values of dispersive and polar components of γ_1, γ_2, and γ_w. Body 1 refers to the polymer surface and body 2 to the adherent plasma protein. In Table 1 γ_1 and γ_1^d are listed as reported in literature. γ_1^p can be calculated from $\gamma_1 - \gamma_1^d$. The γ_1 values for cellulose and poly(vinyl alcohol) (PVA) have not yet been determined; hence, they were calculated on the assumption of the validity of the geometric mean rule for γ_1^p. γ_{1w} is also given in Table 1. γ_w^d and γ_w^p used

Table 1. Surface free energies and work of adhesion of various polymers with bovine serum albumin in water (unit in erg $\cdot$ cm^{-2})

	γ_1	γ_1^d	γ_{1w}	$W_{12,\,w}$
Polyethylene (HD) [40]	35.0	35.0	52.6	1.40
Polyethylene (HD) [41]	34.7	34.7	52.5	1.37
Polyethylene (LD) [42]	33.2	33.2	52.2	1.20
Polypropylene [42]	28.5	28.5	51.4	0.72
Polytetrafluoroethylene [40]	19.5	19.5	51.2	0.25
Paraffin [40]	25.5	25.5	51.1	0.40
Polytrifluoroethylene [41]	23.4	23.4	51.0	0.17
Polystyrene [42]	42.0	41.4	43.7	3.21
Polystyrene [41]	44.8	43.6	40.3	3.82
Nylon 11 [43]	44.0	43.1	41.9	3.59
Nylon 11 [41]	43.8	42.8	41.2	3.63
Poly(vinyl chloride) [42]	41.5	40.0	37.7	3.60
Poly(ethylene terephthalate) [41]	43.9	41.3	33.7	4.10
Poly(ethylene terephthalate) [42]	47.3	43.2	29.8	4.59
Poly(vinylidene chloride) [42]	45.0	42.0	32.5	4.26
Poly(methyl methacrylate) [42]	40.2	35.9	27.4	3.88
Nylon 6,6 [42]	47.0	40.8	24.6	4.58
Poly(vinyl fluoride) [42]	36.7	31.3	24.1	3.54
Polyurethane-E [44]	69.9	47.6	10.8	4.91
Polyurethane-T [44]	63.4	43.7	11.1	4.69
Polyurethane-S [44]	52.0	29.3	6.20	3.04
Cellulose [45]	59	30	3.7	2.7
Poly(vinyl alcohol) [45]	59.5	29	3.1	2.5
Canine artery [46]	70.9	20.9	0.02	0.18
Canine vein [46]	70.6	21.4	0.02	0.19
Water	72.8	21.8	0	0
(Air)	0	0	72.8	0

$\gamma_1^p = \gamma_1 - \gamma_1^d$

for the calculation are 21.8 and 51.0 erg $\cdot$ cm^{-2}, respectively [41]. γ_{1w} is a quantitative measure of the polymer hydrophilicity; the material surface with higher γ_{1w} is more hydrophobic and that with $\gamma_{1w} = 0$ is completely hydrophilic.

For the calculation of $W_{12,w}$ we further need γ_2^d and γ_2^p. In this work, the values for bovine serum albumin (BSA) reported by Paul and Sharma are used, i.e. 31.4 and 33.6 erg $\cdot$ cm^{-2} for γ_2^d and γ_2^p, respectively [47]. Absolutely correct values for these parameters are not necessary, as we merely intend to establish a qualitative relationship between $W_{12,w}$ and γ_{1w}.

Figure 1 shows the plot of $W_{12,w}$ against γ_{1w}. The data are scattered but apparently give a curve with a maximum. The two closed circles refer to imaginary materials, "water" and "air". In both cases, $W_{12,w}$ is zero, γ_{1w} being 0 (water) and 72.8 erg $\cdot$ cm^{-2} (air). Often, proteins are selectively adsorbed on the air/solution interface. The calculations made have nothing to do with this selective adsorption; they only indicate that there is no adhesion between air and protein. For comparison, W_{12} was calculated according to Eq. (4) and the result was plotted in Fig. 1. As anticipated, the work of adhesion in vacuum increases continuously with increasing hydrophilicity of polymers. In contrast, the work of adhesion in water determined under the assumption $[\gamma_{12}]_w = \gamma_{12}$ increase with rising hydrophobicity (the result is not shown here). This trend is in contrast to the general observation and indicates that Eq. (13) is not a correct expression for the work of adhesion in water.

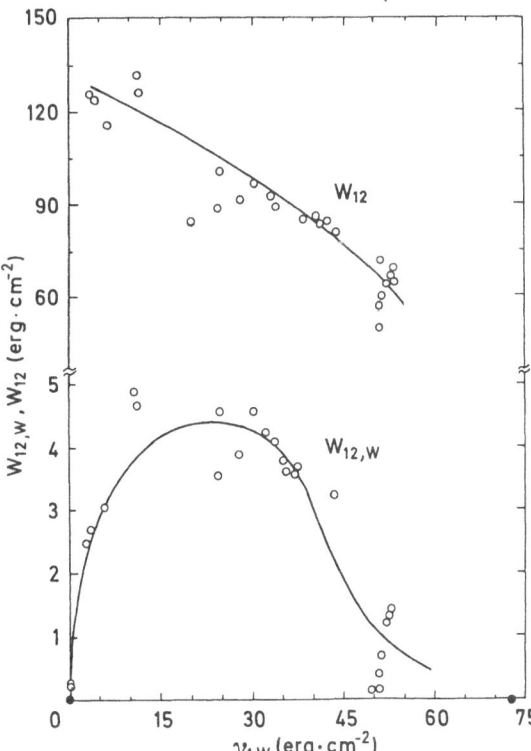

Fig. 1. Work of adhesion of polymer surface 1 with bovine serum albumin 2 in water ($W_{12,w}$) and in vacuum (W_{12}) as a function of the interfacial free energy between water and polymer (γ_{1w})

On the other hand, the estimated dependence of $W_{12,w}$ on γ_{1w} is in good agreement with the result obtained for protein adsorption. The amount of BSA adsorbed onto different films is shown in Fig. 2. The PVA film gives the minimum amount of protein adsorbed. Appreciable differences in the data for polyethylene (PE) and polypropylene (PP) may be explained in terms of surface oxidation which readily occurs in the case of polyolefins. The $W_{12,w}$ values in Fig. 1 for materials with high γ_{1w} (about 50 erg $\cdot$ cm^{-2}) seem to be somewhat too small when we recall that proteins and cells are attached to a considerable extent to silicone and PTFE [48]. This low $W_{12,w}$ may be due mainly to inadequate assumption for the numerical value of γ_2^d and γ_2^p. Although the shape of the curve in Fig. 1 is strongly dependent on γ_2^d and γ_2^p, a maximum always appears in the $W_{12,w}$-γ_{1w} plot, irrespective of the γ_2 value used.

There is a controversy about expressing the work of adhesion in vacuum by Eq. (8) [49]. γ_1^p is usually calculated by means of Eq. (8), without any strong theoretical support since there are only few methods available at present for estimating γ_1^p. Therefore, it should be noted that Eq. (12) is a very rough approximation of the work of adhesion in water. In addition, the curve in Fig. 1 is obtained from unsatisfactory experimental results. Nevertheless, it qualitatively describes the features which have been generally observed. For instance, in the case of highly hydrophilic polymers with very low γ_{1w}, adhesion takes place with difficulty in water, in contrast to adhesion in air, whereas highly hydrophobic polymers with very high γ_{1w} do not undergo adhesion either in air or in water [50]. Strong adhesion is achieved for polar polymers with a medium γ_{1w} like dishes for cell culture [51]. These findings are apparently supported by the result presented in Fig. 2. It follows that, at least qualitatively, the adhesion phenomena in aqueous media can be discussed on the basis of the $W_{12,w}$-γ_{1w} plot in Fig. 1, provided that only dispersive and polar forces are operative in adhesion.

Figure 1 clearly demonstrates that there are two possibilities for a polymer surface to possess a $W_{12,w}$ value equal to zero, in other words, to become non-adhesive. One

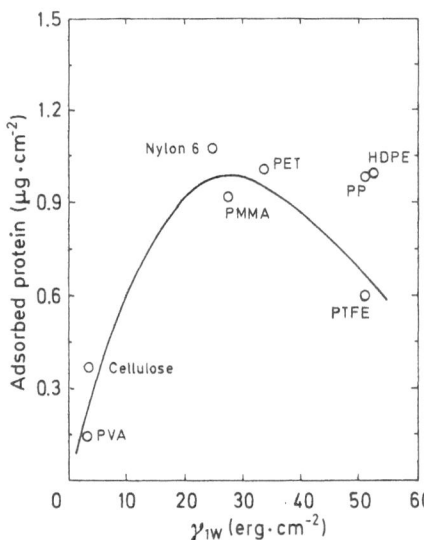

Fig. 2. Bovine serum albumin (BSA) adsorption to different polymer surfaces at 37 °C in phosphate-buffered solution. The initial BSA concentration is 3 mg $\cdot$ ml^{-1}. PVA: poly(vinyl alcohol), PMMA: poly(methyl methacrylate), PET: poly(ethylene terephthalate), PP: polypropylene, HDPE: high-density polyethylene, PTFE: polytetrafluoroethylene

method is to create a *superhydrophilic* surface ($\gamma_{1w} = 0$) while the other is to produce a *superhydrophobic* surface ($\gamma_{1w} = 73$ erg $\cdot$ cm). This answers the old and still debatable question why two extremely different surfaces are relatively blood-compatible.

4.3 Superhydrophilic Diffuse Surface

A hydrophilic surface appears to be more promising for a long-term blood-compatible polymer. The main reason is that it is almost impossible to synthesize a hydrophobic polymer which exhibits a lower $W_{12,w}$ than perfluoropolymers which possess the lowest γ_1 among conventional polymers. However, it may be much easier to modify the polymer surface so as to have a $W_{12,w}$ close to zero. For instance, surface grafting with water-soluble polymers would yield a material the surface of which is covered with water or an aqueous layer. The model structure is schematically shown in Fig. 3. This structure is called *diffuse surface*. Such a material possesses a normal solid surface in the dry state but produces a diffuse interface when brought into contact with an aqueous medium. The diffuse interface may contain a large amount of water and a very small amount of water-soluble polymer chains covalently bound to the substrate material of good mechanical properties. It is likely that this two-phase system has γ_{1w} and $W_{12,w}$ both parameters being very small or almost close to zero if the interfacial volume consists mostly of water. It should be noted that the surface of the so-called hydrogels, which generally contain about 50 wt-% water, must be relatively sharp rather than diffuse even when brought into contact with water.

Grafted soluble
chain

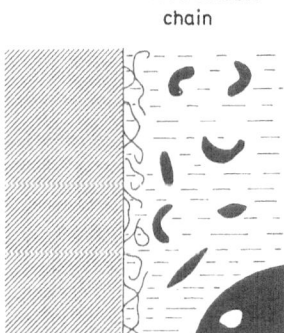

Fig. 3. Schematic representation of the interface between the diffuse surface and blood components

Material   Blood

5 Methods for Preparing Diffuse Polymer Surfaces

Van der Waals interactions occurring in aqueous media between the macromolecular solutes and the polymer surface covered with a thin water layer may be insignificant, as mentioned above. Such interfacial structure can be produced, for instance, by grafting water-soluble chains onto a polymer surface. If a polymer material itself has such chains on its surface, grafting is not necessary.

Although polymer grafting has extensively been studied, there are very few reports on surface grafting [52,53] Surface graft copolymerization of 2-hydroxyethyl methacrylate (HEMA) is the most widely known method of surface grafting. However, this polymerization does not yield a diffuse surface because the HEMA polymer is insoluble in water. Moreover, a diffuse surface with a very high water content is not obtained even for grafting of water-soluble polyacrylamide (PAAm) chains, if crosslinks are introduced among the PAAm chains. The occurrence of grafting not only at the surface but also near and within the polymer substrate is admitted, unless the mechanical properties of the substrate are deteriorated. Important is that the outermost surface is really covered with the grafted chains which are otherwise entirely soluble in water.

There are in principle three methods for surface modification to generate a diffuse structure. 1. Binding of water-soluble polymer chains to the substrate surface like enzyme immobilization [54]. 2. Polymerization of water-soluble monomer by initiating in the substrate surface and propagating toward the outside. 3. Coating of the polymer.

5.1 Graft Coupling Method

A prerequisite of this binding method is that both the polymer surface and the water-soluble polymer chain should contain reactive groups. It seems desirable to use a polymer containing no or only a very small number of ionic groups since the latter must interact electrostatically with the blood components. Therefore, water-soluble polymers with hydroxy groups such as PVA, dextran, and poly(ethylene glycol) are suitable for coupling. Also PAAm and poly(N-vinyl pyrrolidone) can be used if reactive groups are introduced by copolymerization or end-group modification.

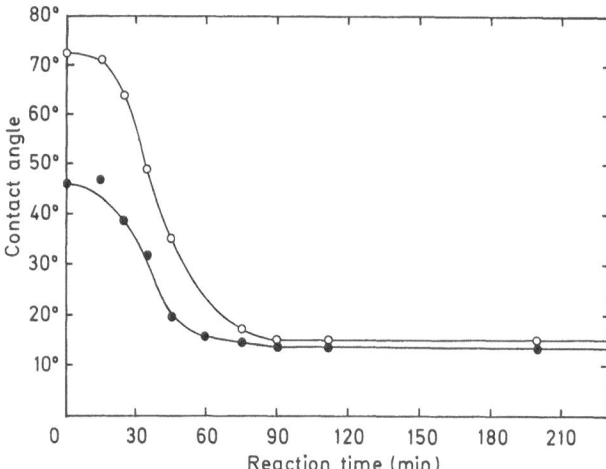

Fig. 4. Effect of duration of urethanation with 10 vol-% hexamethylene diisocyanate in toluene on the contact angles of vinyl alcohol-ethylene copolymer films grafted with dextran for 4 min at polymer concentration of 10 wt-%. ○ advancing; ● receding

In the following the coupling reaction of dextran with a substrate polymer containing hydroxy groups will be described [55]. The reaction scheme is as follows:

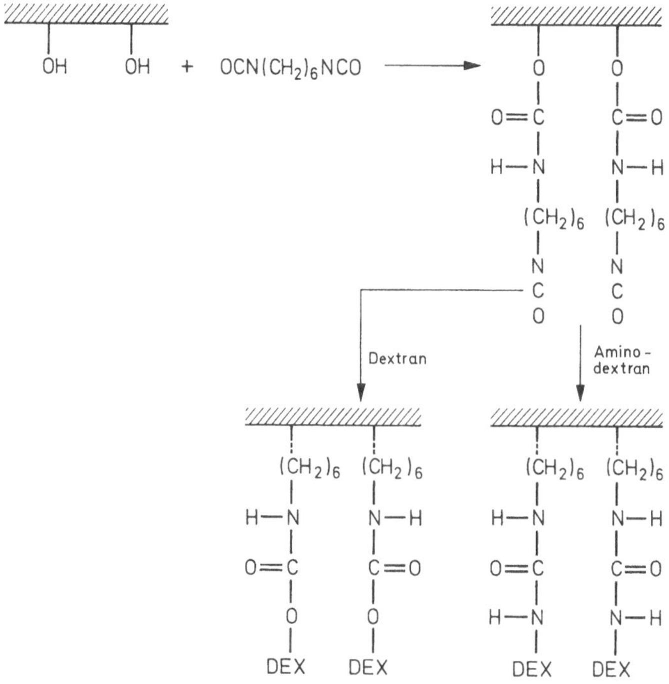

First, alkylene diisocyanate reacts with the polymer surface to introduce isocyanate groups which are capable of reacting with hydroxy and amino groups with the formation of urethane and urea, respectively. Dextran and aminodextran are coupled with the diisocyanate. Figure 4 shows the contact angles of films urethanated with hexamethylene diisocyanate in toluene at different periods followed by grafting with dextran for 4 min in the presence of ferric chloride which serves as a catalyst for urethanation. In this case, a vinyl alcohol-ethylene copolymer (VAECO) film is used since it contains hydroxy groups but is insoluble in water. Moreover, it can be injection-molded to yield a material with excellent mechanical properties. Figure 4 describes the grafting of the hydrophobic surface with the water-soluble chains. Coupling of PVA onto the VAECO film gives a similar result [56]. However, the amount of grafted chains could not be determined, because no efficient method of determination was available. When a radiolabeled gelatin is grafted in a similar manner onto a VAECO film, the amount of gelatin grafted can be determined with a scintillation counter to be approximately $1 \ \mu g \cdot cm^{-2}$ [56].

5.2 Graft Copolymerization Method

Among many other synthetic methods *free-radical polymerization* has almost exclusively been used for surface grafting. To initiate the radical polymerization from a substrate surface, free radicals or peroxides should be generated on the surface. For

this purpose, we can use (1) catalytic initiators such as benzoyl peroxide, azobisiso-
butyronitrile, and ceric ions (2) UV radiation (3) high-energy radiation such as gamma
rays and electron beams (4) glow discharge (5) ozone.

Of these methods, three are briefly described.

5.2.1 Catalytic Polymerization

Free radicals may be produced on polymer chains as a result of hydrogen abstraction
when a chemical initiator is decomposed after mixing with the polymer substrate.
Compared with this method, *graft copolymerization* occurs with a much higher probability if the chemical initiator is covalently bound to the polymer substrate.

An example of the fixation of azo groups at a polymer surface is given by the following reaction [57].

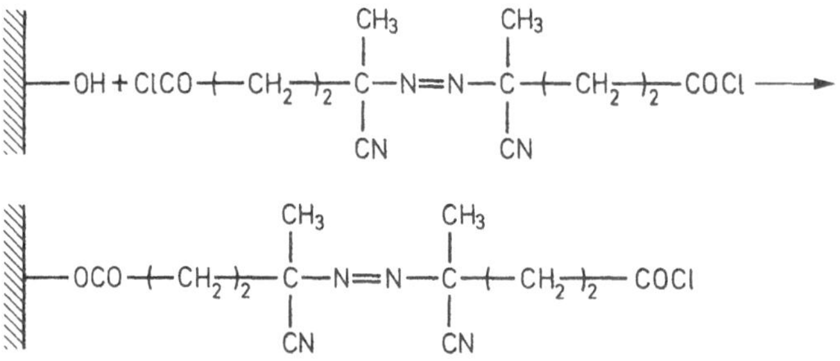

I

Cellulose, PVA, and their derivatives can be used as polymers which have hydroxy
groups on the surface. As azo component the commercially available 4,4'-azobis-4-
cyanovaloyl chloride (AIVC) was employed. Whether or not the azo group is really
introduced into the polymer can be verified, for instance, by further reaction with a
spin-labeled compound like 2,2,6,6-tetramethyl-4-aminopiperidine-1-oxyl (TMPO).
This reaction proceeds as follows:

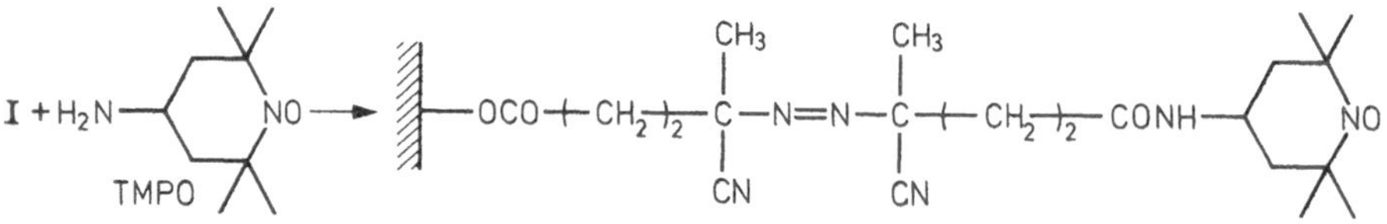

Figure 5 shows the electron spin resonance (ESR) spectroscopy results for the reaction
of cellulose-AIVC-TMPO before and after UV irradiation in toluene for different
durations at room temperature. The reaction of cellulose with AIVC was performed
in dioxan solution because cellulose is insoluble in hydrophobic solvents. Since
dioxan swells the hydrophilic film only weakly, the hydroxy groups located at and
near the film surface are possibly esterified with AIVC. Clearly, the shape of the ESR
spectrum changes upon irradiation, resulting in the appearance of three sharp lines
and the disappearance of a broad line. As is known, the ESR spectrum changes from
a sharp with several lines to a broad one-line shape with increasing radical concentration. This phenomenon is called *spin-exchange broadening effect*. This effect

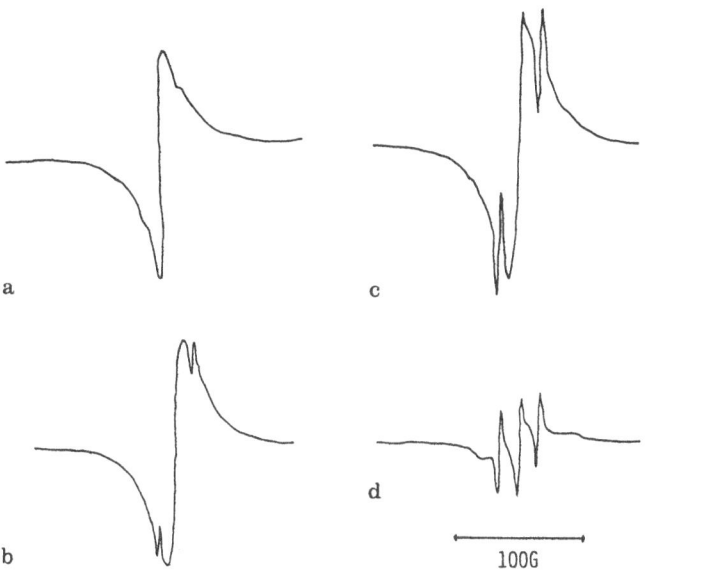

a

c

b

d

100 G

Fig. 5a—d. ESR spectra of cellulose-AIVC-TMPO irradiated by UV in toluene for 0 min (**a**), 25 min (**b**), 85 min (**c**), and 40 h (**d**)

is reduced when the azo bond is cleaved by UV irradiation to liberate the linked TMPO, in other words, to lower the nitroxyl concentration in the film. The free TMPO molecules in toluene exhibit three sharp ESR peaks which are characteristic of free rotation of the molecule.

The ESR study clearly reveals that AIVC really reacts with the hydroxy groups at or near the film surface. A similar result was also obtained for other films such as PVA and VAECO. Graft copolymerization of various monomers onto these films was conducted at 50 °C by immersing them in the liquid monomer. Measurements of contact angles were undertaken to examine whether or not graft copolymerization actually took place onto the film surface. The increase in weight, which is usually determined in conventional graft copolymerization, was too small to be accurately measured. The contact angles of water of PVA-AIVC films grafted by heating are listed in Table 2. Also for the VAECO-AIVC we obtained almost the same result as for PVA film. Table 2 clearly shows that graft copolymerization gives rise to a noticeable change in wettability of the PVA film. The higher contact angle of the PVA-AIVC in comparison with that of the virgin film is probably due to the weak hydrophilicity of the AIVC group even in its hydrolyzed form. As expected, graft copolymerization with acrylic acid and subsequent neutralization with NaOH remarkably reduces the contact angle of the PVA-AIVC film. Since the polymer of 2-(N,N-dimethylamino)ethyl methacrylate (DMAEM) is not soluble in water, the film grafted with this monomer exhibits a contact angle which is not significantly different from that of PVA-AIVC. It is not surprising that graft copolymerization with N-vinyl pyrrolidone (N-VP) results in an increase of the contact angle, although N-VP polymer is soluble in water, because the hydrogel prepared by irradiation of

Table 2. Contact angles of water on poly(vinyl alcohol) grafted with various monomers

Monomers	Contact angle (deg.)	
	Receding	Advancing
virgin	27.5	36.6
AIVC (hydrolyzed)	36.1	55.6
Acrylic acid	21.8	31.9
Sodium acrylate	11.8	13.1
2-(N,N-Dimethylamino)ethyl methacrylate	35.2	46.5
N-Vinyl pyrrolidone	41.5	57.1
Methyl methacrylate	49.6	61.0

the aqueous solution of this polymer with electron beams shows a large contact angle with respect to water. The contact angle of the PVA-AIVC film grafted with methyl methacrylate (MMA) is almost the same as that of the MMA homopolymer.

5.2.2 Radiation Polymerization

Most studies of *radiation-induced graft copolymerization* result from attempts to produce a polymer material which possesses both good mechanical properties and a hydrogel-like surface [58-62]. As mentioned above, the hydrogels do not always have a diffuse surface. PHEMA is strongly swollen in water, but not soluble in water and has a relatively large contact angle. Even for water-soluble polymers, the introduction of crosslinks reduces the diffuse nature of the surface. It is not easy to produce a diffuse layer only at the outermost surface of a biomaterial. However, it does not matter whether graft copolymerization proceeds in the interior of the material, if the mechanical properties of the material are not greatly deteriorated by grafting.

The preparation of a diffuse polymer surface by radiation-induced graft copolymerization is schematically described in Fig. 6. When a polymeric material is subjected to high-energy radiation, free radicals are formed in the material which subsequently react with oxygen, if present, to form peroxides. Their production is probably not restricted to the surface region [Fig. 6 (I)]. When the irradiated material is put in a deaerated monomer solution containing an adequate amount of reducing agent [Fig. 6 (II)], the peroxides are reduced to radicals even at room temperature due to the very low activation energy for the redox reaction [Fig. 6 (III)]. If the reducing agent is present in excess, the radicals undergo deactivation. In the absence of a reducing agent, the peroxide bond can be thermally cleaved to radicals, but in this case thermal polymerization of the monomer may occur simultaneously to a considerable extent. The addition of monomer to the radicals is followed by chain propagation as shown in Fig. 6 (IV). Polymerization would proceed even within the material if the monomer diffusion is enhanced, for instance, by raising the temperature or adding a swelling agent. In addition to the diffusion of monomer into the virgin substrate, the inner concentration of active species capable of initiating polymerization may increase the possibility of polymerization in the interior since grafted chains, once formed inside the material, would promote monomer diffusion.

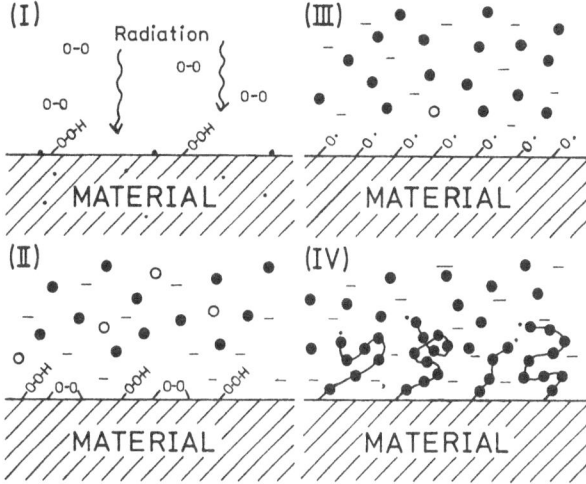

Fig. 6. Schematic representation of the surface graft copolymerization induced by radiation. ● monomer; ○ reducing agent

In the following, some of the experimental results are given supporting the above mechanism of the graft copolymerization of acrylamide (AAm) in the absence or presence of a reducing agent (FeCl$_2$) onto PE films of high density (HDPE) and low density (LDPE) pre-irradiated in air [63, 64]. If graft copolymerization is carried out at 50 °C in the absence of Fe^{2+}, the wettability by water greatly increases without weight increase, as shown in Fig. 7 and Table 3. On the other hand, graft copolymerization carried out at 30 °C in the presence of a certain amount of Fe^{2+}, leads to an unexpectedly large increase in weight. The apparent graft yield (in percent), defined as [(weight of graft film/weight of starting film)-1] × 100, amounted to several thousand percent. Such a large weight increase can also be observed for the graft copolymerization carried out at 50 °C without Fe^{2+} by pre-irradiation technique in vacuum.

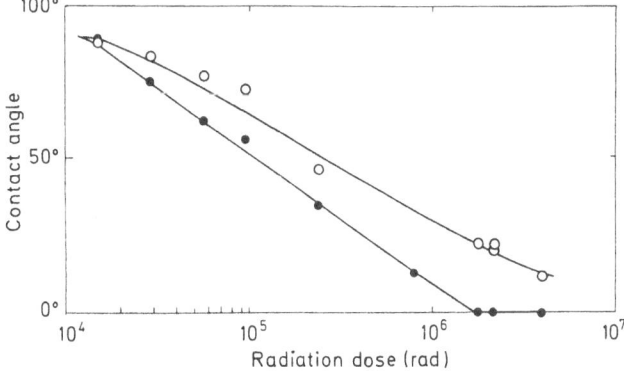

Fig. 7. Dependence of the contact angle of water on the preirradiation dose. Acrylamide was polymerized with a 10 wt-% solution for 1 h at 50 °C in the absence of Fe^{2+}. ○ low-density polyethylene; ● high-density polyethylene

Table 3. Weight and contact angle change of polyethylene tubes grafted by copolymerization of acrylamide for 90 min (pre-irradiation in air; dose 2.4×10^5 rad)

		Polyethylene	
		Low density	High density
Weight	before graft copolym.	10.217	1.582
(mg)	after graft copolym.	10.220	1.570
	before pre-irradiation	90	90
Contact	after pre-irrad. but		
angle	before graft copolym.	84	80
(deg.)	after graft copolym.	17	0

In order to modify the substrate surface without altering any bulk property, the weight increase associated with a significant change in size should be suppressed as much as possible. When polymerization is carried out in the presence of Fe^{2+} at 15 °C, practically neither any weight increase nor dimensional change is observed. In addition, the PE film becomes wettable by graft copolymerization if the Fe^{2+} concentration is kept within a certain range. The plot of the contact angle as a function of the Fe^{2+} concentration is given in Fig. 8. The remarkable feature is the decrease in contact angle which, after passing through a minimum, increases again with higher Fe^{2+} concentration. Homopolymer is formed to an insignificant extent and the grafted films from HDPE are rendered more wettable than those from LDPE, unless the radiation dose is very high. Graft copolymerization at 30 °C in the presence of Fe^{2+} for the LDPE film gives rise to a higher weight increase than for HDPE. These results indicate that the phase structure of the PE film affects graft copolymerization.

Similar results to the graft copolymerization by pre-irradiation in air of PE with gamma rays from Co-60 were obtained for that by pre-irradiation in air with electron beams from a van de Graaff accelerator.

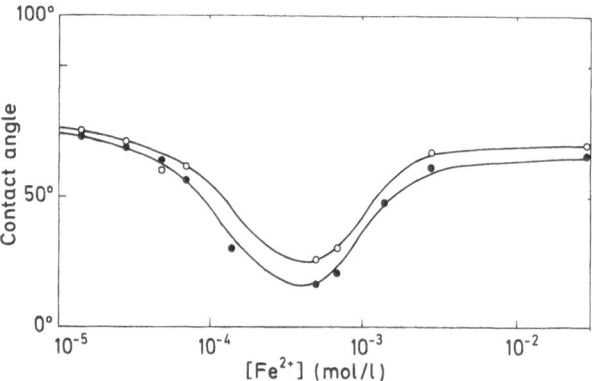

Fig. 8. Dependence of the contact angle of water on Fe^{2+} concentration (pre-irradiation; dose 1.0×10^6 rad; polymerization of acrylamide in a 10 wt-% solution for 57 h at 15 °C). O low-density polyethylene; ● high-density polyethylene

5.2.3 Plasma-Induced Polymerization

The non-equilibrated, low-temperature plasma which is generated by glow discharge in gases is very suitable for the modification of polymer surfaces. Examples are *plasma polymerization* and *plasma treatment* [65]. In comparison with these plasma applications, *plasma-induced graft copolymerization* has attracted the attention of only very few researchers.

In the so-called plasma polymerization, the gaseous monomer is converted to an extraordinarily densely crosslinked polymer which is grafted to the substrate or simply deposits on it. On the other hand, the plasma-induced graft copolymerization is similar to radiation graft copolymerization using the pre-irradiation technique. In these cases, the active species for polymerization (free radicals or peroxides) are generated separately before the polymerization of monomer starts. If the polymeric material subjected to plasma or high-energy radiation is not exposed to an oxygen-containing atmosphere prior to polymerization, free radicals will initiate directly graft polymerization. On the other hand, graft copolymerization starts following decomposition of peroxides if the substrate polymer is irradiated in air or exposed to air immediately after irradiation.

The oxygen-free plasma-induced graft copolymerization with free radicals was extensively studied by Fales and his coworkers [66], primarily in order to improve the dyeability of fibers, whereas very little is known about the graft copolymerization initiated by peroxide formed from oxygen [67]. The plasma-induced graft copolymerization via peroxide formation is briefly described below [68].

The plasma generation apparatus used in our laboratory is equipped with a plate rotated by a motor. The specimen to be exposed to the plasma is fastened to the plate. Since the frequency is 5 kHz, a matching unit is not required. Figure 9 describes the amount of peroxides decomposed upon heating at 70 °C (for further details see the legend to this figure). The amount was determined from the change of the 1,1-diphenyl-2-picrylhydrazyl concentration in benzene solution in which the plasma-treated films were immersed. If it is assumed that peroxide decomposition is virtually completed by heating for about 5 h, the amount of the peroxides formed by plasma treatment may be close to 1.5×10^4 peroxy groups cm^{-2}. If a grafted chain having a molecular

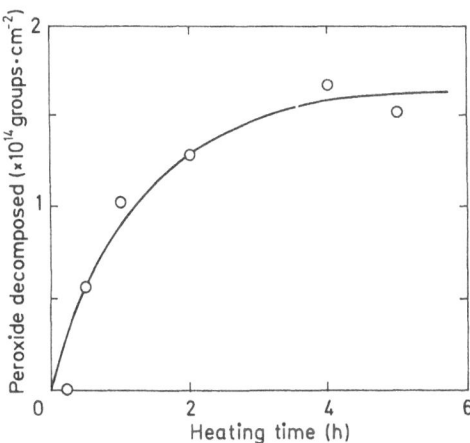

Fig. 9. Concentration of peroxides decomposed by heating at 70 °C in benzene. High-density polyethylene films were pretreated with argon plasma for 5 s

weight of, say, 10^5 is assumed to propagate from all the peroxides, the weight of total grafted chains amounts to 25 µg · cm^{-2}.

The effect of the duration of plasma treatment is described in Fig. 10 for grafting of AAm onto HDPE films pretreated with argon plasma. Polymerization was carried out in a 10 wt-% solution of AAm at 50 °C for 2 hr. The amount of grafted PAAm

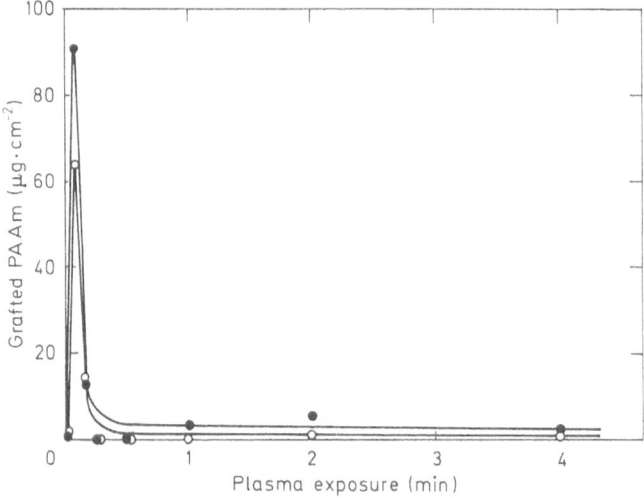

Fig. 10. Influence of argon plasma exposure time on graft copolymerization. Acrylamide was grafted onto plasma-treated polyethylene films in 10 wt-% solution at 50 °C for 2 h. ○ high-density polyethylene; ● low-density polyethylene

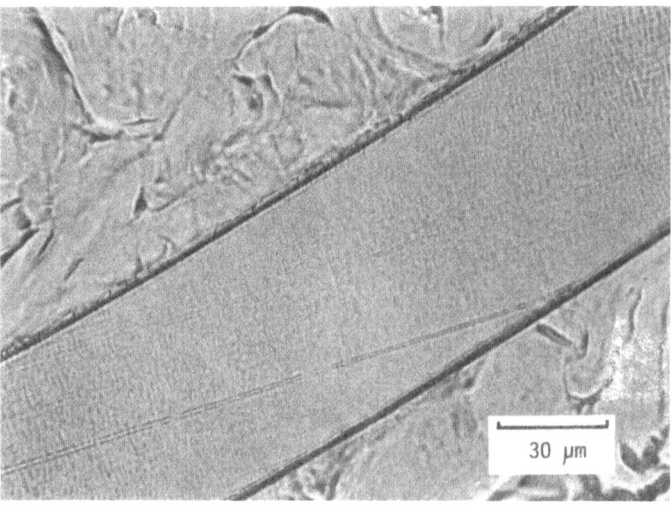

Fig. 11. Light-microscopic photograph of a cross section of an acrylamide-grafted polyethylene film with a graft amount of 90 µg cm^{-2}. Dyeing with toludine blue 0 was performed after alkaline hydrolysis of polyacrylamide to polyacrylic acid

was determined by the nynhidrin method after complete hydrolysis of PAAm. As Fig. 10 reveals, there is an optimum period of plasma treatment for graft copolymerization. The value of the maximum treatment period is reproducible; however no reasonable explanation for the strong effect of the treatment time on graft copolymerization has been reported so far. A similar result has been obtained for plasma treatment in air.

Figure 11 shows the photograph of the cross section of the AAm- grafted PE film. It can be seen that the depth of the grafted layer is about 1 μm. The water wettability of the PE film is improved even by plasma treatment alone but gradually decreases on standing in air. In contrast, such return to the poor wettability of the starting PE is no more observed if the plasma-treated film is further grafted with AAm. The result is represented in Fig. 12.

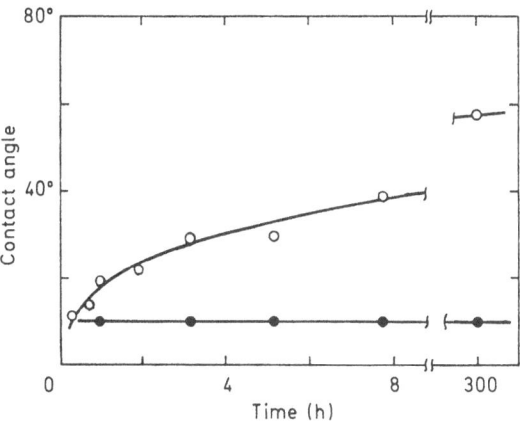

Fig. 12. Change of the contact angle of water on standing in air at room temperature for high-density polyethylene treated with argon plasma followed by graft copolymerization with acrylamide. ○ before grafting; ● after grafting

5.3 Coating Method

The simplest technique for modifying the surface of polymer is to coat the surface with a thin layer of an adequate polymer. In general, crosslinks are introduced into the coated polymer to prevent the coated layer from being dissolved in the surrounding media. However, for the formation of a diffuse surface, excess crosslinking must be avoided because the diffuse nature of the surface diminishes with crosslinking. The main disadvantage of the coating method is the lack of reliable adhesion between the thin layer and the substrate. If a strong bonding can be achieved by anchoring the layer in a porous substrate, an interesting diffuse surface may be prepared by coating with weakly "crosslinked" biopolymers such as gelatin and mucopolysaccharides. It is possible that the "biolyzed" surface[10] belongs to this category.

6 Surface Analysis of Diffuse Structures

Obviously, a diffuse structure is formed only if the surface is in contact with an aqueous medium. Otherwise, a normal surface structure as sharp as other hydrophilic materials like glass, oxidized PE, cellulose, and PHEMA would result. Therefore, to examine whether a material exhibits a diffuse structure, a surface analysis should be performed under such a condition as the surface is in contact with water. It has been pointed out that gel-like biomaterials used in contact with blood should also be analyzed when in contact with water [69]. If so, physical means such as XPS, Auger spectroscopy, and scanning electron microscopy (SEM) are not so powerful for the surface characterization of biomaterials. Unfortunately, there are very few methods available for the analysis of the diffuse structure.

6.1 XPS(ESCA)

As mentioned above, we cannot obtain a decisive evidence for the presence of a diffuse structure from the XPS analysis alone. However, in some cases, XPS provides strong support for the diffuse structure. For instance, if the XPS result on the elemental analysis is nearly identical with the elemental analysis value of the homopolymer of the grafted monomer, we may conclude that the surface region of the material is covered with the grafted polymer chains. This is the case with the graft copolymerization of AAm on PE films [64]. XPS result of the films pre-irradiated in air followed by polymerization of AAm, gives the spectrum characteristic of N_{1s} which is due to PAAm. The elemental ratios calculated from the XPS spectra are listed in Table 4 which also compiles the results for the starting PE film and the AAm homopolymer. It is clearly seen that the O/C and N/C ratios of the AAm-grafted films are identical with those of the AAm homopolymer within experimental error. These findings indicate that exclusively the PAAm chains must occupy the surface region, at least, of several tens Å depth. Thus, it is likely that the grafted films form a diffuse interface when brought into contact with water since water-soluble chains are grafted without any crosslinking agent. However, this analysis provides us no information on the water content of the diffuse layer.

Table 4. Comparison of XPS elemental ratios for polyethylene, polyacrylamide, and acrylamide-grafted polyethylene

	Wt. increase (%)	Elemental ratio	
		O/C	N/C
HDPE	—	0.02	0
LDPE	—	0.01	0
AAm-grafted HDPE[a]	≈ 0	0.23	0.21
AAm-grafted LDPE[b]	≈ 0	0.26	0.22
AAm-grafted HDPE[c]	1400	0.26	0.24
PAAm	—	0.24	0.24

HDPE: high-density polyethylene; LDPE: low-density polyethylene;
AAm: acrylamide;
a) pre-irradiation (dose 2.1×10^6 rad), polymerization at 50 °C;
b) pre-irradiation (dose 2.1×10^6 rad), polymerization at 50 °C;
c) pre-irradiation in vacuum (dose 1.3×10^6 rad), polymerization at 50 °C

6.2 Zeta-Potential

A very powerful analytical method for studying the interface with water is to measure
the zeta-potential which is related to electrokinetic phenomena. Briefly, the zeta-
potential is the electric potential at the slipping plane or the plane of shear located at
the zone from which ions are displaced by the flowing electrolyte solution. The surface
in contact with the electrolyte is believed to determine which ions are released from the
slipping plane. In addition, the zeta-potential is closely related to the potential of the
surface which contains chemically adsorbed ions [70].

There are several methods for determining the zeta-potential. Generally, the streaming potential measurements have been used for the characterization of biomaterials
being in contact with an aqueous environment [71]. Figure 13 describes the pH dependence of the zeta-potential for uncoated and siliconized glass surfaces with adsorbed
BSA [72]. The results for the glasses before adsorption of BSA are clearly different from
those in Fig. 13 (see Fig. 15). The isoelectric point estimated from the plots is 4.8 in both
cases, which is in good agreement with that of native BSA. This agreement indicates
that zeta-potential measurements afford valuable information on the interface with
water. Further examples of zeta-potential measurements are given in Figs. 14 and 15.
The ionic strength of the flowing electrolyte solution is 0.001. The samples in Fig. 14
are PVA films grafted with various monomers [57]. As is seen from Fig. 14, the effect
of surface modification can be clearly recognized from zeta-potential measurements.
For instance, positive values of the zeta-potential, irrespective of pH, indicate that a

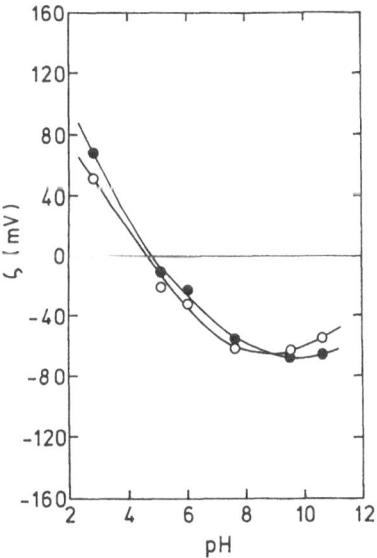
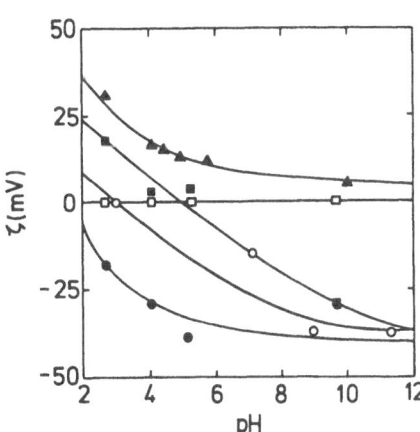

Fig. 13. pH Dependence of zeta-potential for surfaces adsorbed with bovine serum albumin. ○ glass;
● siliconized glass

Fig. 14. pH Dependence of zeta-potential for poly(vinyl alcohol) grafted with different monomers.
□ before grafting; ■ *N*-vinyl pyrrolidone; ○ methyl methacrylate; ● acrylic acid; ▲ 2-(*N*,*N*-di-
methylamino)ethyl methacrylate

Yoshito Ikada

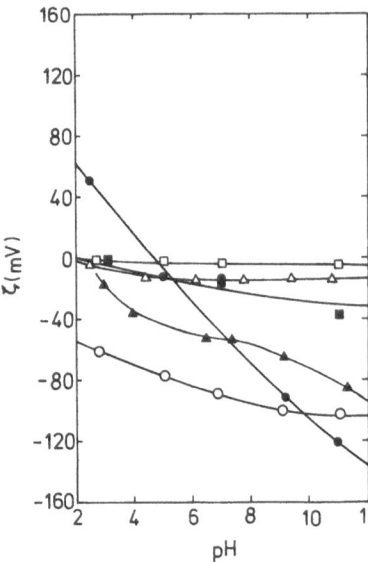

Fig. 15. pH Dependence of zeta-potential for various surfaces. ○ glass; ● siliconized glass; □ PVA not annealed; ■ PVA annealed in N_2 for 15 min at 210 °C; △ cellulose; ▲ vinyl alcohol-ethylene copolymer

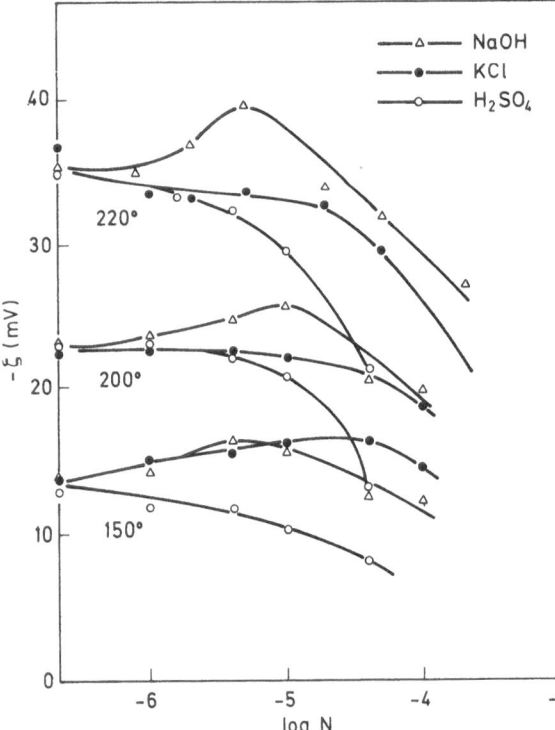

Fig. 16. Zeta-potential of poly-(vinyl alcohol) fibers heat-treated at 150 and 220 °C. Streaming potential was measured in NaOH, KCl, and H_2SO_4 at different concentrations (N) (O. Yoshizaki[73])

cationic monomer, DMAEM, has actually been graft-copolymerized onto the PVA surface. Although both MMA and N-VP polymers are nonionic, the variation of their zeta-potential with pH is different. This may be attributed to the difference in the adsorption of ionic species on the polymer surfaces. The zeta-potentials of various nonionic polymers are shown in Fig. 15 [72]. As is seen, they are virtually zero or negative over the pH range studied for all surfaces except the siliconized surface. PE has higher negative potentials than VAECO, which, in turn, has larger negative values than PVA. Zeta-potentials of hydrophilic surfaces such as PVA and cellulose are close to zero. Clearly, heat treatment of PVA causes a shift to the zeta-potentials to higher negative values. A similar result was also reported by Yoshizaki [73], as shown in Fig. 16.

The result for PE grafted with AAm and arylic acid (AA) is shown in Fig. 17 [72]. It is obvious that graft copolymerization of AAm onto PE reduces the zeta-potential to zero over the whole pH range studied. This pH dependence of the AAm-grafted PE is the same as that of the PVA which is not heated.

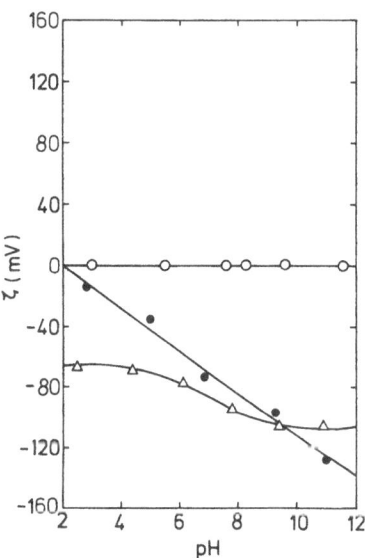

Fig. 17. pH Dependence of zeta potential for polyethylene grafted with water-soluble monomers. ● before grafting; ○ acrylamide; △ acrylic acid

6.3 Protein Adsorption

In the studies on the blood-compatible polymers, protein adsorption is very important at least for two reasons. One reason is triggering of thrombus formation by protein adsorption as mentioned above. The other reason is that protein adsorption reflects the structure of the interface between the material and water, thus providing a good method for the analysis of the surface being in contact with water.

Figure 18 shows the amount of BSA adsorbed by PVA films plotted as a function of their equilibrated water content [72]. The films with different water contents were prepared by changing the temperature of the heat treatment both in air and N₂. It is seen that the heat treatment in air results in higher protein adsorption than in N₂,

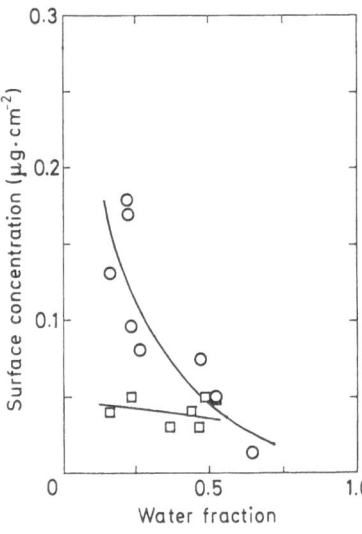

Fig. 18. Influence of the water fraction of poly(vinyl alcohol) on bovine serum albumin adsorption. ○ annealed in air; □ annealed in N$_2$

suggesting that oxidation which has taken place during the heat treatment in air increases the work of adhesion in water (see Fig. 1). Clearly, protein adsorption becomes less prominent with increasing water content of the bulk films. Combining this result with that of the zeta-potentials of PVA, we can propose a model for the structure of the interface between PVA and water (Fig. 19) [72]. Atactic PVA is a crystalline polymer and normally insoluble in water below about 80 °C. Heat treatment of dried PVA films increases the crystallinity and consequently decreases the degree of swelling in water at room temperature. Therefore, it seems highly possible that, as depicted in Fig. 19, some chains on the surface of thermally untreated PVA are not involved in the crystalline region but are extruded towards the outer water phase to give a diffuse surface. The soluble chains fixed at the surface shift the slipping plane to the outer phase causing an extremely low zeta-potential.

The interface model given in Fig. 19 is very similar to that of the surface grafted with a monolayer of water-soluble heterochains as illustrated in Fig. 3. Practically, it is very difficult to control the thickness of grafted chains, because grafting proceeds not only on the substrate surface but also near the surface [74]. In this connection, an interesting behavior during protein adsorption was observed in the case of VAECO films grafted with PVA and dextran by the coupling method. Fig. 20 shows the results for the adsorption of BSA and fibrinogen [56]. As is clearly seen, adsorption on the dextran-grafted films increases continuously for both proteins as the grafting proceeds, whereas adsorption on the PVA-grafted films decreases first with grafting, passes through a minimum and then increases gradually until saturation. Such a large amount of proteins adsorbed on the dextran-grafted films cannot be explained in terms of conventional surface adsorption. It seems that interactions between solutes and the substrate interior must also be taken into account. Popov and his coworkers have revealed that BSA is *sorbed* by hydrogels prepared from crosslinked PHEMA and PAAm (Fig. 21) [75].

The difference in results observed in Fig. 20 between PVA and dextran grafting is not yet clearly understood but their adsorption behavior may be explained as follows.

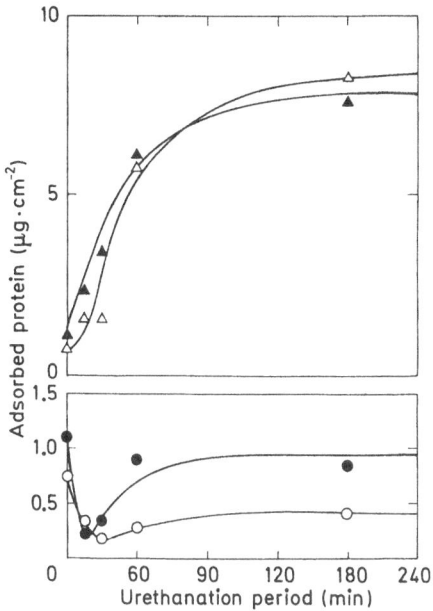

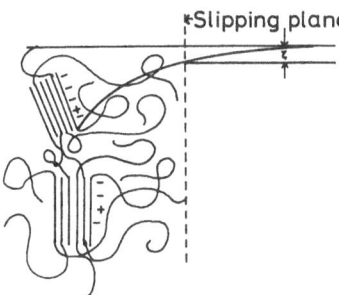

Fig. 19. Model of the electric double layer formed on the surface of poly-(vinyl alcohol) being in contact with water

Fig. 20. Adsorption of bovine serum albumin (BSA) and fibrinogen on vinyl alcohol-ethylene co-polymer films coupled with poly(vinyl alcohol) (lower part) and with dextran (upper part). Adsorption was carried out at 37 °C for 3 h with BSA (3 mg · ml^{-1}) and fibrinogen (0.45 mg · ml^{-1}). ○ PVA, BSA; ● PVA, fibrinogen; △ dextran, BSA; ▲ dextran, fibrinogen

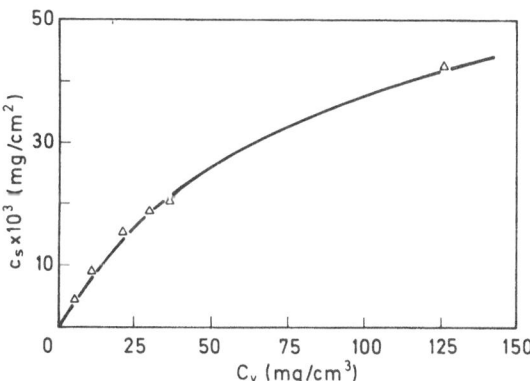

Fig. 21. Isothermal curve of the *sorption* of serum albumin-[131] I at 37 °C with a pH of 7.4 for poly-acrylamide gel films containing 60 % water and having an average thickness of ~ 100 μm (K. N. Popov et al. [75])

When VAECO is grafted with dextran, the layer becomes very thick; this leads to high *sorption* of proteins. On the other hand, in the grafting with PVA, coupling takes place predominantly on the surface to give a diffuse structure. As a consequence, protein adsorption decreases at the beginning of grafting. However, the interaction between the protein and the grafted chains becomes appreciable as the grafting proceeds. A model of the interaction is given in Fig. 22 [76]. It is likely that the interaction is weaker than that between the protein and a solid surface. The latter adsorption is accompanied by protein denaturation which may result in thrombus formation.

The results of protein adsorption on PE films grafted with AAm by radiation polymerization are given in Fig. 23 [77]. As protein, fibronectin is used which is known to play an important role in cell adhesion [78]. A reduction in protein adsorption by graft copolymerization is obvious.

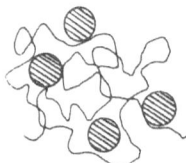

Fig. 22. Mutual exclusion of polymer molecules and spherical particles (J. Hermans [76])

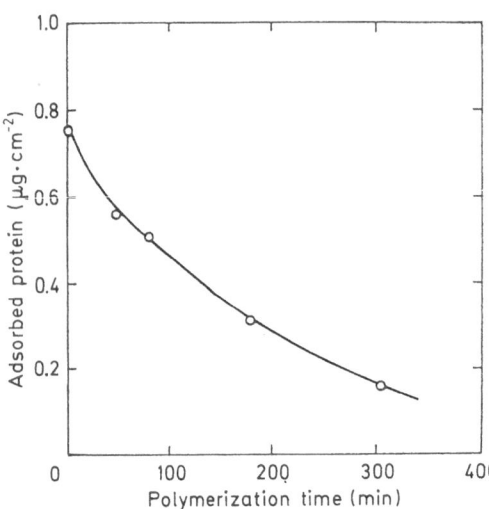

Fig. 23. Adsorption of fibronectin onto polyethylene grafted with acrylamide by the radiation method. Adsorption was performed in 0.05 mg · ml^{-1} protein solution

7 Evaluation of Blood Compatibility

A large obstacle to the rapid development of the blood-compatible polymers is the lack of an evaluation standard for blood compatibility. This lack of standard method is, to some extent, unavoidable since thrombus formation is a combination of such complex events that no single method can fully evaluate the compatibility of bio-materials. A variety of tests should be accomplished taking into account the hydro-dynamics of blood, the physical and chemical structure of the interface between the blood and the biomaterial, the sort of test animals, etc.

7.1 *In Vitro* Test with Platelets

The polymer surface undergoes platelet adhesion immediately after interacting with plasma proteins when brought into contact with blood. Therefore, platelet adhesion has most often been examined in the studies on blood compatibility of polymers. Figures 24 shows the phase-contrast microscopic photographs of various surfaces brought into contact with platelet-rich plasma (PRP) [72]. The photographs were taken of the surfaces onto which the PRP-containing ACD (acid citrate dextrose) was placed at 37 °C for 5 min. Small black spots in the photographs are platelets; an inspection reveals that platelet adhesion to the surfaces decreases in the following order:

AA-grafted PE > glass > PE > siliconized glass > PVA > sodium acrylate (AANa)-grafted PE

When $CaCl_2$ is added to the PRP, fibrin deposition occurs on some of the surfaces. The amount of fibrin formed decreases in the following order:

glass > PE > siliconized glass > AAm-grafted PE

AA-grafted PE does not give rise to any fibrin formation, although numerous platelets are adherent to the polymer surface. Interestingly, neither platelet adhesion nor fibrin formation is noticeable on the surface of AANa-grafted PE. This finding is in marked contrast to AA-grafted PE on which many platelets are deposited irrespective of $CaCl_2$ addition. Such behavior must be associated with the fact that a white precipitate is formed when PRP is mixed with a dilute aqueous acidic solution of poly(acrylic acid) (PAA). For AAm-grafted PE and PVA, only few platelets are deposited and fibrin formation is not observed.

7.2 *Ex Vivo* Test with Whole Blood

Clot formation in polymer-coated glass tubes was examined by introducing canine whole blood directly into the tubes according to the method illustrated in Fig. 25 [72]. The amount of clots was estimated by means of the kinetic method of Komai and Nosé [79]. The results for different polymers are given in Fig. 26. As can be seen, uncoated glass exhibits the lowest and PVA-coated material the highest thromboresistance. The order of the blood compatibility of these materials, evaluated by this method, is apparently the same as that obtained by the test with PRP. The very insignificant but still detectable clotting occurring on the PVA surface indicates that PVA is not completely resistant to thrombus formation. However, it should be also remembered that in this evaluation technique the blood in the tube does not flow at all. To study in more detail the clotting on PVA, heat treatment of PVA in N_2 was carried out at different temperatures. Figure 27 shows that the blood compatibility decreases due to heat treatment [72]. SEM of the surface being in contact with the whole blood for 5 min reveals no trace of clot on the thermally untreated PVA surface, whereas fibrin deposition has occurred on the uncoated glass surface and partly on other surfaces. This result is almost similar to the observation by light microscopy of the surfaces brought into contact with PRP.

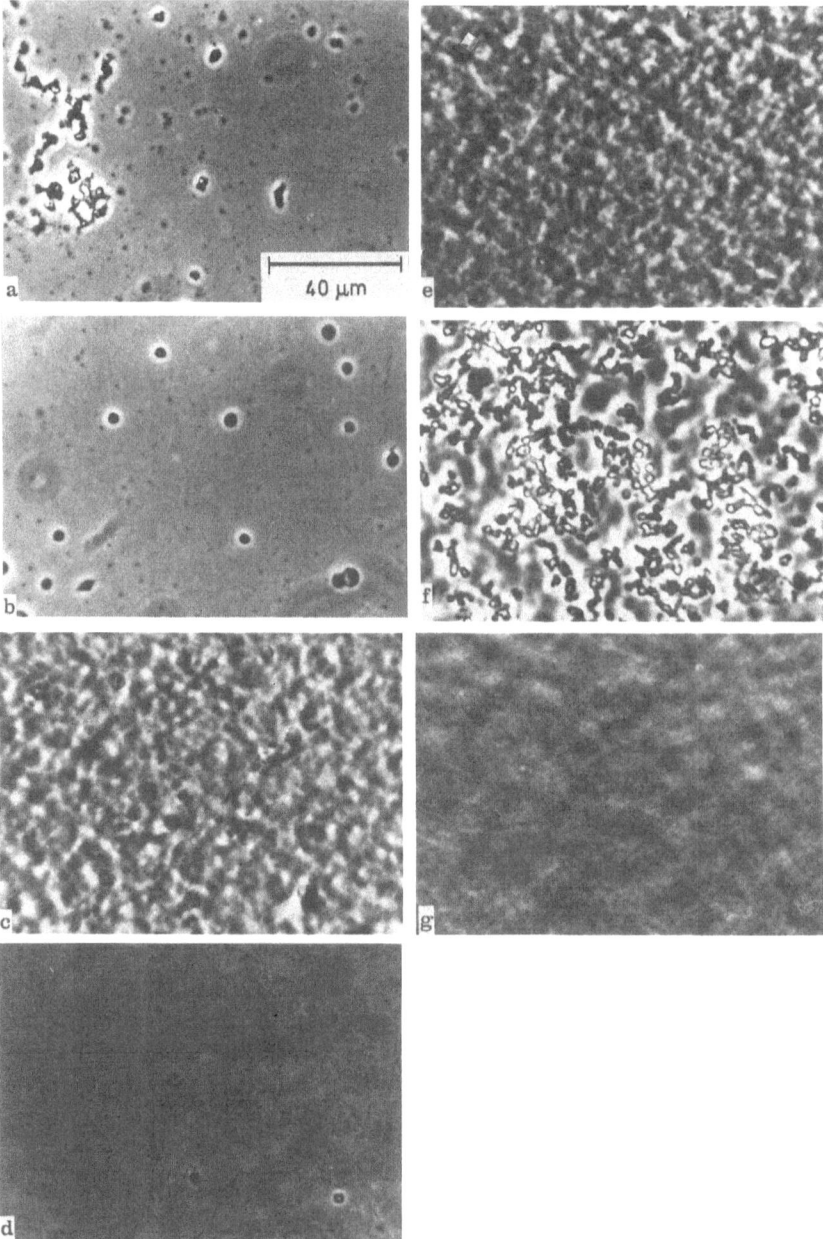

Fig. 24a—g. Phase-contrast micrographs of surfaces which are in contact with platelet-rich plasma. **a**, glass; **b**, siliconized glass; **c**, polyethylene; **d**, poly(vinyl alcohol); **e**, acrylamide-grafted polyethylene; **f**, acrylic acid-grafted polyethylene; **g**, sodium acrylate-grafted polyethylene. Scale is given in a

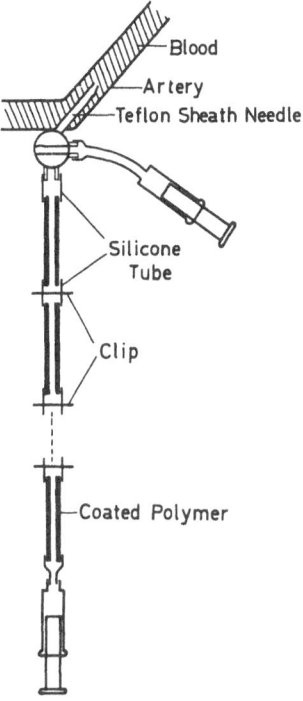

Fig. 25. Experimental arrangement for filling the fresh blood into polymer-coated glass tubes

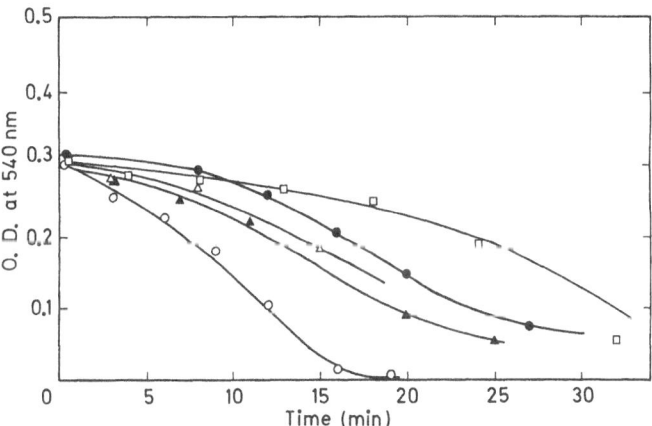

Fig. 26. *Ex vivo* evaluation of the blood compatibility of various surfaces. ○ glass; ● siliconized glass; □ poly(vinyl alcohol); △ vinyl alcohol-ethylene copolymer; ▲ polyethylene

Ex vivo evaluation by kinetic methods has revealed that the coupling grafting with dextran onto VAECO films improves the blood compatibility to some extent [55].

7.3 *In Vivo* Test Using Anastomotic Method

Evaluation *in vivo* is essential for the blood-compatible polymer, because the final purpose is the clinical use. Thus, various methods have been proposed for *in vivo*

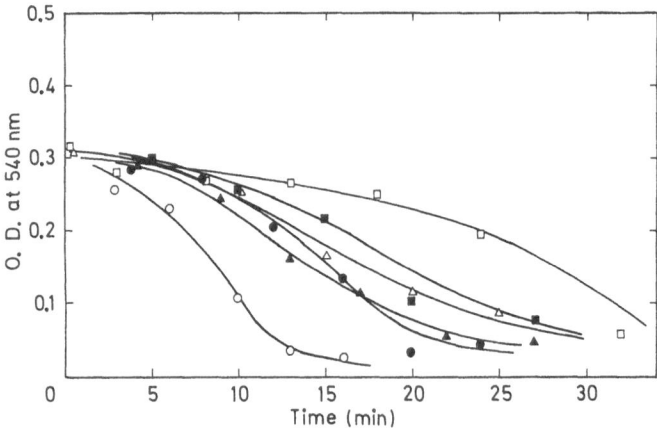

Fig. 27. *Ex vivo* evaluation of the blood compatibility of poly(vinyl alcohol) annealed in N_2 at different temperatures. □ 60 °C; ■ 150 °C; △ 170 °C; ▲ 210 °C; ○ glass; ● siliconized glass

evaluation of the blood compatibility. They consist either of the insertion of fibers, sheets, and rings into blood vessels or of the replacement of a part of the arteries or veins by tubes whose surface consists of the polymer to be tested. An arteriovenous shunt is also used for testing the blood compatibility of tubes. We employed an anastomotic method in which tubes are connected with common carotids of rat.

The patency results of PE tubes grafted with AAm and AA by radiation are summarized in Table 5 [80]. The length and the inner and outer diameter of the tubes are 1.8 cm, 0.8 mm, and 1.0 mm, respectively. Anastomosis of the tubes with the carotid was performed using a cyanoacrylate adhesive and a polymer splint soluble in blood [81]. The results are compiled in Table 5.

As Table 5 reveals, occlusion takes place within 20–30 min with the formation of clots when the ungrafted PE tubes are anastomosed with the rat carotid. Graft copolymerization with AA is virtually not able to prolong the occlusion time. In contrast to the PRP experiment, neutralization of AA chains with NaOH does not improve the blood compatibility when the test is performed with flowing blood. On the other

Table 5. Patency results[a] for different tubes (I.D. 0.8–1.0 mm, length 18 mm, common carotid of rats, non-suture anastomosis)

	Duration of patency	Occlusion due to
PVA[b]	7–28 day	disconnection of anastomotic region
AAm[b]-grafted PE[b]	1–7 day	disconnection of anastomotic region
AA[b]-grafted PE	20–30 min	thrombus formation
PE	15–30 min	thrombus formation

a) average of 5 rats for each tube;
b) PVA: poly(vinyl alcohol); AAm: acrylamide; PE: polyethylene; AA: acrylic acid

hand, grafting with AAm greatly enhances the blood compatibility of the PE tube as in the case of the test with PRP. Occlusion occurs 1 to 7 days after anastomosis, but is not due to clotting on the tube surface but disconnection of the tubes from the carotids. This disconnection mostly results from the large difference in the Young's modulus of PE and the carotid although a soft LDPE was used as the starting tube. Even for the tubes grafted with AAm, poor patency is observed if small grooves are detected by SEM on the surface. The importance of the surface smoothness and matching of the tube compliance has also been pointed out by Pierce [82] and Annis [83], respectively.

Instead of the hard PE tube which may exert an untolerable stress on the pulsating soft carotid, the patency rate of PVA tubes which were not thermally treated and hence very soft have been studied. The obtained results are summerized in Table 6 [80]. The length and the outer and inner diameter of this tube are 1.8 cm, 0.8–0.9 mm, and 1.1–1.2 mm, respectively. Anastomosis was not carried out with the cyanoacrylate adhesive but with a suture of nylon 11-0 since the adhesive was biodegradable, leading to disconnection of the tube from the carotid. The cyanoacrylate adhesive was only used in the case where bleeding from needle holes of the PVA tube did not cease be-

Table 6. Patency results for poly(vinyl alcohol) tube (I.D. 0.8–0.9 mm, O.D. 1.1–1.2 mm, length 18 mm, common carotid of rats, anastomosed with an 11-0 nylon suture)

No. of rats	Patency after 1 week		Patency after 1 month	
	No. of patent tubes	%	No. of patent tubes	%
10	8	80	7	70
10	—	—	7	70

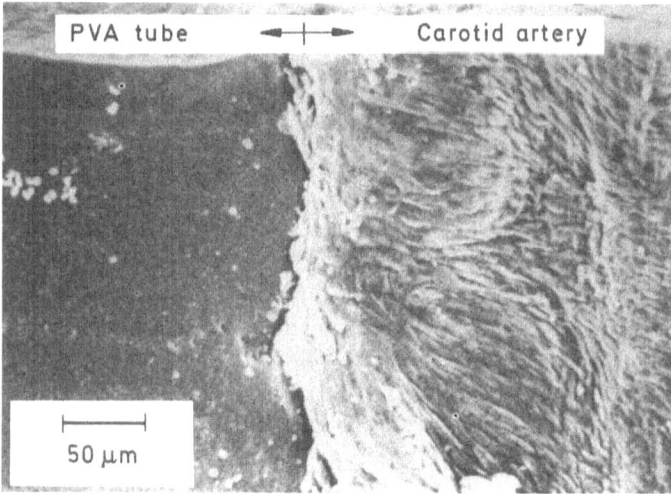

Fig. 28. SEM photograph of the inner surface of a poly(vinyl alcohol) tube connected to a rat carotid artery 4 weeks after implantation

cause of the good blood compatibility of PVA. The anastomosed PVA pulsated together with the carotid artery.

As Table 6 shows, eight of ten tubes were patent when inspected one week after operation. Two tubes were occluded because of bleeding from the suture line. Inspection after one month revealed that seven of eight tubes, which were not occluded at the first inspection, and seven of ten which were not examined at the time of the first inspection — totally fourteen of twenty tubes — were patent at the second inspection. Hematoma was observed in the connected region of four occluded tubes. Half of the occluded tubes were disconnected from the carotid. Connective tissues covered the outer surface of all the tubes, but no thrombus was formed on the inner surface of all the occluded and patent tubes. Neither aneurysmal dilation nor burst was observed in the anastomotic region.

Figure 28 shows an SEM photograph of the anastomotic region four weeks after operation. As is seen, the end of the carotid is covered with endothelium cells while the PVA surface is virtually free of any cells, indicating that the PVA surface is entirely antithrombogenic at least within the test period.

8 Conclusion

In this article it is shown that a completely antithrombogenic surface can be obtained from synthetic polymers without any help of bioactive polymers as heparin and urokinase. Such a surface has a *diffuse structure*, which essentially differs from that of so-called hydrogels which have a relatively low water content. There is no reason to suspect that the interaction of the polymer surface with plasma proteins initiates a series of complex biochemical events leading to thrombus formation. Recently it has been reported that even if proteins deposit on a material to a multilayer, the proteins at the outermost layer of multi-layered protein deposit might remain intact, which would eventually prevent platelet adhesion [84].

The starting point of our work is to find a polymer surface that only slightly interacts with the plasma proteins to prevent the initiation reaction. Therefore, we have theoretically derived the work of adhesion in water and represented graphically this parameter as a function of the free energy of the interface formed between the polymer surface and water, referring to physicochemical data reported in the literatures. The theoretical derivation includes several assumptions partly because the surface chemistry of polymers is still in its infancy. The physical parameters necessary for estimating the work of adhesion in water have not yet been accumulated to a sufficient extent. Qualitatively, however, our calculations led to the conclusion that blood-compatible polymers possess either an extremely hydrophobic or extremely hydrophilic surface. We have focused our attention on the hydrophilic surface since extremely hydrophobic surfaces cannot be prepared at present.

An extremely hydrophilic surface is readily obtained by coating a polymer surface with a mobile layer of water. A monolayer from bound water is not sufficient since such a monolayer would interact with proteins to a comparatively significant extent. The mobile, relatively thick water layer is formed on the surface which contains water-soluble polymer chains covalently or firmly attached to the surface. Since these chains are capable of keeping a large quantity of water inside and around them, a mobile water layer is spontaneously formed on the surface when it comes into contact with water. The water-soluble chains grafted onto the surface should contain no ionogenic groups since the latter would be involved in electrostatic interactions with plasma proteins.

The grafted surface exhibits not a sharp but a diffuse boundary surface when coming into contact with water. It is likely that cell membranes with oligosaccharide chains on the surface show such a diffuse interface, but we do not intend to mimic the surface structure of endothelium cells because they presumably acquire the outstanding blood compatibility by virtue of bioactive substances such as prostacyclin [85]. Slippery surfaces of mucous membranes must also possess a diffuse structure. Similarly, the diffuse surface in question will hardly stimulate the flowing blood; in other words, it will not significantly interact with plasma proteins. Therefore, this surface can be regarded as a non-stimulating bio-inert surface. As is well-known, non-ionic water-soluble polymers such as poly(N-vinyl pyrrolidone), poly(ethylene glycol), PAAm, and dextran do not cause thrombus formation if injected into the flowing blood. Thus, it is reasonable to assume that the surface grafted with these water-soluble polymers may be blood-compatible. PAAm gels used for electrophoresis and dextran gels used for gel permeation of proteins are known to adsorb an insignificant amount of proteins unless these gels are strongly crosslinked and charged [86].

We have not yet accumulated the sufficient experimental results which allow us to discuss details on the physical structure of the diffuse surface. It has not been determined so far how thick and dense the diffuse layer must be to attain good blood compatibility, partly because no powerful analytical methods for determining the length and concentration of the grafted chains on the surface are available at present.

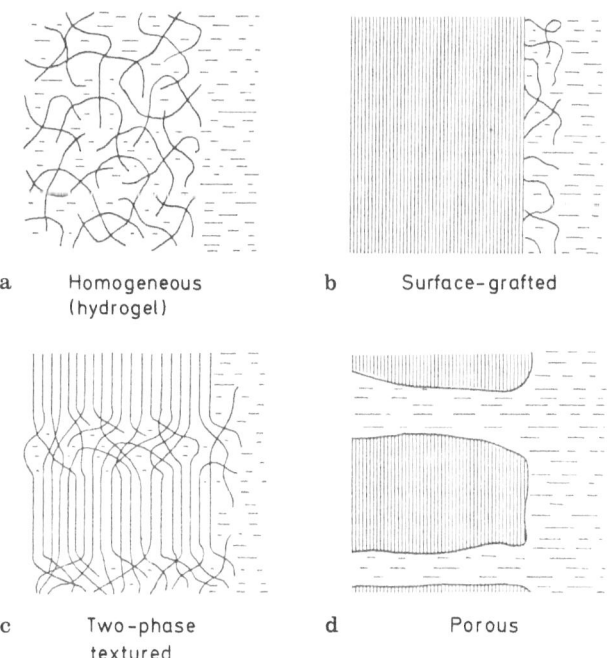

a   Homogeneous   b   Surface-grafted
       (hydrogel)

c   Two-phase   d   Porous
       textured

Fig. 29 a—d. Models of surface in contact with water and having very low interfacial free energy

Even the water content in the diffuse layer cannot be determined with sufficient accuracy. Furthermore, it should be noted that the true surface modified by the above-mentioned methods is much more complex than the simple model depicted in Fig. 3. Practically, it is almost certain that grafting takes place not only on the surface but also below the polymer surface which is much more permeable to chemical reagents than inorganic materials like hard glass. Consequently, protein *sorption* occurs occasionally in addition to surface adsorption, resulting in complicated inter- actions with the polymeric material.

There still remain many problems to be solved for developing an excellently blood- compatible polymer. At least, a direct comparison of the blood compatibility *in vivo* by means of a powerful evaluation method is required for each of the polymers which different research groups have synthesized according to their own hypothesis. Our diffuse surface hypothesis is only one of these hypotheses. The materials illustrated in Fig. 29 may contain a large fraction of water on their surface, leading to minimum interactions with blood.

Finally, it should be stressed that an antithrombogenic surface is merely one con- dition of the blood compatibility of biomaterials which can be used in medicine. If the biomaterial is implanted as blood pumps of artificial heart and vascular grafts, we should also take into consideration other important properties such as mechanical durability, calcification, bonding strength with respect to host tissue, etc. Studies on blood-compatible polymers have been started only since recently.

Acknowledgement. The auther is pleased to acknowledge discussions and colla- boration with Dr. H. Iwata of Research Institute of National Cardiovascular Center, Osaka, Dr. F. Horii of Institute for Chemical Research, Kyoto University and Drs. H. Handa, Y. Yonekawa, W. Taki, S. Yamagata, and H. Miyake of Department of Neurosurgery, Kyoto University.

9 References

1. Ulrich, H., Bonk, H. W., Colovos, G. C.: Synthetic Biomedical Polymers; Szycher, M., Robinson, W. J. (Ed.) Technomic Publ. Co., Connecticut, p. 29, 1980
2. Ratner, B. D.: Photon, Electron, and Ion Probes of Polymer Structure and Properties; Dwight, D. W., Fabish, T. J., Thomas, H. R. (Eds.) ACS Symp. Series *162*, 371 (1981)
3. Costa, V. S. D., Brier-Russel, D., Salzman, E. W., Merrill, E. W.: J. Colloid Interface Sci. *80*, 445 (1981)
4. Chawla, A. S.: J. Biomed. Mater. Res. *16*, 501 (1982)
5. Graham, S. W., Hercules, D. M.: J. Biomed. Mater. Res. *15*, 349 (1981)
6. Matsumoto, H., Kimura, T., Takamatsu, T., Fukada, E.: Jinkozoki *5*, 152 (1976)
7. Andrade, J. D. (Ed.) Hydrogels for Medical and Related Applications, ACS Symp. Series *31*, 1976
8. Wichterle, O., Lim, D.: Nature *185*, 117 (1960)
9. Iwamoto, G. K., King, R. N., Andrade, J. D.: Photon, Electron, and Ion Probes of Polymer Structure and Properties; Dwight, D. W., Fabish, T. J., Thomas, H. R. (Eds.) ACS Symp. Series *162*, 405 (1981)
10. Nosé, Y., Tajima, K., Imai, Y., Klain, M., Mrava, G., Schriber, K., Urbanek, K., Ogawa, H.: Trans. Am. Soc. Artif. Intern. Organs *17*, 482 (1971)
11. Salzman, E. W., Lindon, J., Brier, D., Merrill, E. W.: Ann. N.Y. Acad. Sci. *283* 114 (1977)
12. Bair, R. E., Akers, C. K.: Trans. Am. Soc. Artif. Intern. Organs *27*, 770 (1978)

13. Ward, R. S., Tsuchiyama, B. J., O'Connor, K. A.: Trans. Am. Soc. Artif. Intern. Organs 27, 410 (1980)
14. Hecher, J. F., Edwards, R. O.: J. Biomed. Mater. Res. 15, 1 (1981)
15. Bair, R. E., Gott, V. L., Feruse, A.: Trans. Am. Soc. Artif. Intern. Organs 16, 50 (1970)
16. Schrader, M. D.: J. Colloid Interface Sci. 88, 296 (1982)
17. Andrade, J. D.: Med. Instrum. 7, 110 (1973)
18. Coleman, D. L., Gregonis, D. E., Andrade, J. D.: J. Biomed. Mater. Res. 16, 381 (1982)
19. Ratner, B. D., Hoffman, A. S., Hanson, S. R., Harker, L. A., Whiffew, J. D.: J. Polym. Sci., Polym. Symp. 66, 363 (1979)
20. Hanson, S. R., Harker, L. A., Ratner, B. D., Hoffman, A. S.: J. Lab. Clin. Med. 95, 289 (1980)
21. Mahmud, N. A., Wan, S., Sa da Costa, V., Vitale, V., Brier-Russell, D., Kuchner, L., Salzman, E. W., and Merrill, E. W.: Physicochemical Aspects of Polymer Surfaces, Vol. 2; Mittal, K. L. (Ed.) Plenum Press, New York, p. 953, 1983
22. Sawyer, P. N., Srinivasan, S.: Am. J. Surg. 114, 42 (1967)
23. Bruck, S. D.: Nature 243, 416 (1973)
24. Okano, T. Nishiyama, S., Shinohara, I., Akaike, T., Sakurai, Y., Kataoka, K., Tsuruta, T.: J. Biomed. Mater. Res. 15, 393 (1981)
25. Kim, S. W., Lee, E. S.: J. Polym. Sci., Polym. Symp. 66, 429 (1979)
26. Vroman, L., Leonard, E. F. (Eds.): The Behavior of Blood and its Components at Interfaces, Ann. N.Y. Aca. Sci. 283, (1977)
27. Bruck, S. D.: J. Polym. Sci., Polym. Symp. 66, 283 (1979)
28. Salzman, E. W. (Ed.): Interaction of the Blood with Natural and Artificial Surfaces, Marcel Dekker, Inc. New York and Basel, 1981
29. Biomaterials: Interfacial Phenomena and Applications, ACS Adv. Chem. 199, 1982
30. Bantijes, A.: British Polym. J. 10, 267 (1978)
31. Griffin, J. H., Cochrane, C. G.: Semin. Thromb. Hemostasis 5, 254 (1974)
32. Vroman, L., Adams, A. L.: J. Biomed. Mater. Res. 3, 43 (1969)
33. Hoffman, A. S.: ACS Organic Coatings and Applied Polymer Science Proceedings, 48, 28 (1983)
34. Ikada, Y., Suzuki, M., Tamada, Y.: ACS Polymer Preprints, 24(1) (1983)
35. van Oss, C. J.: Ann. Rev. Microbiol. 32, 19 (1978)
36. Absolom, D. R., Francis, D. E., Zingg, W., van Oss, C. J., Neuman, A. W.: J. Coll. Interface Sci. 85, 168 (1982)
37. Gerson, D. F., Scheer, D.: Biochem. Biophys. Acta, 602, 506 (1980)
38. Dexter, S. C.: J. Coll. Interface Sci. 70, 346 (1979)
39. Cherry, B. W.: Polymer Surfaces, Cambridge Univ. Press 1981
40. Fowkes, F. M.: J. Phys. Chem. 66, 382 (1962); 67, 2538 (1963); ACS Adv. Chem. 43, 99 (1964); Ind. Eng. Chem. 56, No. 12, 40 (1964)
41. Ikada, Y., Matsunaga, T.: J. Adhesion Soc. Japan 15, 91 (1979)
42. Owens, D. K., Wendt, R. C.: J. Appl. Polym. Sci.: 13, 1741 (1969)
43. Andrews, R. H., Kinloch, A. J.: Proc. Roy. Soc. London A 332, 385 (1973)
44. Lelah, M. D., Stafford, R. J., Lambrecht, L. K., Young, B. R., Cooper, S. L.: Trans. Am. Soc. Artif. Intern. Organs 27, 504 (1981)
45. Matsunaga, T., Ikada, Y.: J. Colloid Interface Sci. 84, 8 (1981)
46. Yano, U., Komai, T., Kawasaki, T., Hujiwara, Y.: Prep. Japan Soc. Biomater. 2, 45 (1980)
47. Paul, L., Sharma, C. P.: J. Colloid Interface Sci. 84, 546 (1981)
48. Bornzin, G. A., Miller, I. F.: J. Colloid Interface Sci. 86, 539 (1982)
49. Fowkes, F. M., Mostafa, M. A.: Ind. Eng. Chem. Prod. Res. Dev. 17, 3 (1978)
50. Sello, S. B., Stevens, C. V.: Pure & Appl. Chem. 53, 2211 (1981)
51. Grinnell, F.: Intern. Rev. Cytology 53, 65 (1979)
52. Kronick, P. L.: Synthetic Biomedical Polymers; Szycher, M., Robinson, W. J. (Eds.) Technomic Publ. Co., Connecticut, p. 153, 1980
53. Ratner, B. D.: J. Biomed. Mater. Res. 14, 665 (1980)
54. Ikada, Y., Iwata, H., Mita, T., Nagaoka, S.: J. Biomed. Mater. Res. 13, 607 (1979)
55. Taniguchi, M., Samal, R. K., Suzuki, M., Iwata, H., Ikada, Y.: Graft Copolymerization of Lignocellulosic Fibers; Hon, D. N.-S (Ed.) ACS Symp. Series 187, 217 (1982)
56. Ikada, Y., Lonikar, S. V., Ham, M.-S, Tamada, Y.: Polym. Prep. Japan 31, 1785 (1982)

57. Samal, R. K., Iwata, H., Ikada, Y.: Physicochemical Aspects of Polymer Surfaces: Mittal, K. L. (Ed.) Plenum, New York, Vol. 2, 801, 1983

58. Fischer, J. P., Becker, U., Halasz, S.-P., Mück, K.-F., Püschner, H., Rösinger, S., Schmidt, A., Suhr, H. H.: J. Polym. Sci. Polym. Sym. *66*, 443 (1979)

59. Hoffman, A. S.: Radiat. Phys. Chem. *18*, 323 (1981)

60. Ellinghorst, G., Jansen, B.: Radiat. Phys. Chem. *18*, 1111 (1981)

61. Chapiro, A., Domurado, D., Foex-Millequant, M., Jendrychowska-Bonamour, A. M.: Radiat. Phys. Chem. *18*, 1200 (1981)

62. Hayashi, K., Murata, K., Yamamoto, J., Yamashita, I.: Prep. Annual Meeting, Japan Soc. Biomater. *4*, 115 (1982)

63. Ikada, Y., Suzuki, M., Taniguchi, M., Iwata, H., Taki, W., Miyake, H., Yonekawa, Y., Handa, H.: Radiat. Phys. Chem. *18*, 1207 (1981)

64. Suzuki, M., Tamada, Y., Iwata, H., Ikada, Y.: Physicochemical Aspects of Polymer Surfaces; Mittal, K. L. (Ed.) Plenum, New York, Vol. 2, 923 (1983)

65. H. V. Boenig, Plasma Science and Technology, Cornell Univ. Press, Ithaca and London, 1982

66. Fales, J. D., Bradley, A., Howe, R. E.: Vacuum Technology, March, 53 (1976)

67. Kronick, P.: contract No. NIH-NO1-HV-1-2017, Annual Report, July 1976

68. Suzuki, M., Piao, D.-S., Ikada, Y.: Polym. Prep. Japan *31*, 286 (1982)

69. Andrade, J. D., King, R. N., Gregonis, D. E.: Hydrogels for Medical and Related Applications; Andrade, J. D. (Ed.) ACS Symp. Series *31*, 206 (1976)

70. Benes, P., Paulenova, M.: Colloid and Polym. Sci. *251*, 766 (1973)

71. Van Wagenen, R. A., Coleman, D. L., King, R. N., Trioro, P., Brostrom, L., Smith, L. M., Gregonis, D. E., Andrade, J. D.: J. Colloid Interface Sci. *84*, 155 (1981)

72. Ikada, Y., Iwata, H., Horii, F., Matsunaga, T., Taniguchi, M., Suzuki, M., Taki, W., Yamagata, S., Yonekawa, Y., Handa, H.: J. Biomed. Mater. Res. *15*, 697 (1981)

73. Yoshizaki, O.: Kobunshi Kagaku *12*, 414 (1955)

74. Ratner, B. D., Weathersby, P. K., Hoffman, A. S., Kelly, M. A., Scharpen, L. H.: J. Appl. Polym. Sci. *22*, 643 (1978)

75. Popov, K. N., Degterev, I. A., Zaikov, G. Ye.: Polym. Sci. U.S.S.R. *22*, 1790 (1980)

76. Hermans, J.: J. Chem. Phys. *77*, 2193 (1982)

77. Ikada, Y., Suzuki, M., Tamada, Y.: Prep. Regional Meeting of The Chemical Society of Japan (Chubu) *13*, 104 (1982)

78. Hughes, R. C., Pena, S. D., Clark, J., Dourmashkin, R. R.: Exp. Cell. Res. *121*, 307 (1979)

79. Komai, T., Nosé, Y.: Jinkozoki *4*, 114 (1975)

80. Miyake, H., Taki, W., Yonekawa, Y., Handa, H., Suzuki, M., Ikada, Y.: Presented at Annual Meeting of Vascular Syst., Soc. Japan, Gifu, Nov. 23, 1982

81. Yamagata, S., Handa, H., Taki, W., Yonekawa, Y., Ikada, Y., Iwata, H.: J. Microsurgery *1*, 208 (1979)

82. Pierce, W. S.: Polymers in Medicine and Surgery: Kronenthal, R. L., Oser, Z., Martin, E. (Eds.) Polymer Sci. Technology, Vol. 8, 263, 1975

83. Annis, D.: British Polym. J. *10*, 238 (1978)

84. Matsuda, T., Akutsu, T.: ACS Organic Coatings and Applied Polymer Science Proceedings, *48*, 647 (1983)

85. Moncada, S., Gryglewski, R., Bunting, S., Vane, J. R.: Nature *263*, 663 (1976)

86. Porath, J.: Biochem. Soc. Trans. *7*, 1197 (1979)

K. Dušek (Editor)
Received May 27, 1983

Ionizing Radiation and Gas Plasma (or Glow) Discharge Treatments for Preparation of Novel Polymeric Biomaterials

Allan S. Hoffman
Center for Bioengineering and Department of Chemical Engineering,
University of Washington, Seattle, WA. 98195, U.S.A.

There is a wide variety of materials which are foreign to the body and which are used in contact with body fluids. These materials are called biomaterials.

By far, the most diverse use of biomaterials exists within the polymer class: in these organic materials, ionizing radiation and plasma discharges have a unique ability to initiate free radical and ionic reactions without the need to add catalysts or to heat.

There are three basic processes which are utilized for preparing new or modified polymeric biomaterials. They are: (1) surface modification via radiation graft copolymerization or plasma gas discharge; (2) radiation polymerization of pure monomer(s) in solution, or an emulsion, or in the solid state (e.g., below T_G); and (3) radiation crosslinking, in a solution or swollen state, or in the solid state.

Simultaneous or subsequent chemical or biochemical-processing steps can yield novel biomaterials having specific biological activity. Immobilization of enzymes, antibodies, drugs, cells, etc. on or within the radiation-processed material can yield novel biomaterial systems with great potential in the clinic or clinical laboratory. These processes and products are reviewed in this chapter.

1 Introduction

There is a wide variety of materials which are foreign to the body and which are used in contact with body fluids. These materials are called biomaterials. They include polymers (fibers, rubbers, molded plastics, emulsions, coatings, fluids, etc.), metals, ceramics, carbons, reconstituted or specially treated natural tissues, and composites made from various combinations of such materials (Table 1). Some are needed only for short-term applications while others are, hopefully, useful for the lifetime of the individual. Applications include devices or implants for diagnosis or therapy.

Synthetic polymers make up by far the broadest and most diverse class of biomaterials [1, 2]. This is mainly because synthetic polymers are available with such a wide variety of compositions and properties and also because they may be fabricated readily into complex shapes and structures. In addition, their surfaces may be readily modified physically, chemically, or biochemically. Such modifications can have significant influences on biologic responses to the biomaterials. When a foreign biomaterial contacts blood or tissue fluids, the first measurable response in the initial seconds to

Table 1. Classes and Forms of Biomaterials[1]

Classes	Forms

I) Polymers
    A) Fibers
    B) Rubbers
    C) Plastics
	Films or membranes
	Fibers or fabrics
	Tubes
	Powders or particles
	Molded shapes
	Bags or containers, etc.
	Liquids → solids (adhesives)

II) Metals	Cast or molded shapes Powders or particles Fibers
III) Ceramics	Molded shapes Powders or particles Liquids → solids (adhesives)
IV) Carbons	Machined shapes Coatings Fibers
V) Natural Tissues	Fibers Natural forms Also, reconstituted as films, tubes, fibers, etc.
VI) Composites	Coatings Fibrous felts or sheets Fiber or fabric-reinforced shapes, etc.

minutes is the adsorption of biomolecules — usually proteins. This is followed in the next minutes to hours by cellular interactions — especially platelets (in blood), white blood cells or leukocytes (in blood and tissue fluids), and fibroblasts (in tissue fluids) (Fig. 1).

The important material properties which can influence protein and cell interactions at the biomaterial-biologic interface are listed in Table 2. It is probable that surface composition and topography most strongly influence the composition and organization of the initial adsorbed protein layer [3]. It is this layer which mediates subsequent cellular events at that interface. Thus, a great deal of effort has gone into surface modifications and characterization of polymeric biomaterials.

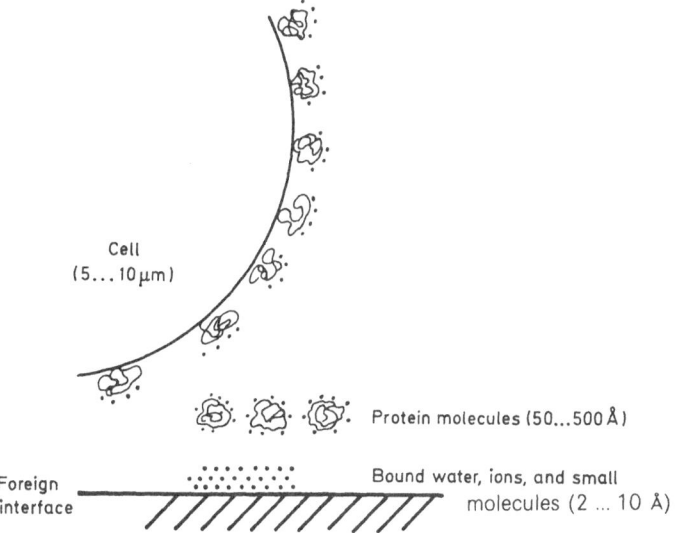

Fig. 1. Interaction of "foreign" polymeric biomaterials with the biologic environment involves protein and cellular interactions at the polymer interface[1]

Table 2. Polymer Material Properties Influencing Biological Responses at Foreign Interfaces [1]

A. Biomaterial Surface
   1. Chemical composition (polar/apolar, acid-base, H-bonding, ionic charges)
   2. Molecular motions (polymer chain ends, loops and their flexibility)
   3. Topography (roughness, porosity, imperfections, gas microbubbles)
   4. Domains (distributions of any of the above in the surface)
B. Initial Interfacial Adsorption
   1. Water ("bound" vs. "structured" vs. "free")
   2. Ions (e. g., calcium ion; acidic or basic ions; local pH; electrochemical reactions)
   3. Lipids
   4. Sugars, Glycosaminoglycans (GAG's)
   5. Proteins (composition and organization)
C. Bulk Absorption, Desorption
   1. Absorption from biologic space (water, ions, lipids, sugars, GAG's, proteins)
   2. Desorption of additives or impurities (plasticizers, stabilizers, unreacted monomers)
   3. Desorption of breakdown products (biodegradation, erosion or corrosion byproducts)

2 Energy Sources for Polymer Surface Modification

Polymer surfaces are commonly modified by physical deposition of other compounds (e. g., surfactants, polymers) by direct chemical modification of the polymer surface (e. g., oxidation, hydrolysis, sulfonation, etc.) or by chemical bonding of a different polymer (graft copolymerization) or "polymerlike" composition (via plasma discharge) on the substrate polymer surface. The most commonly used energy sources for the last two techniques are ionizing radiations [4-7] and radio-frequency or microwave gas discharges, also called plasma or glow discharges [8-10]. Photochemical initiations using U.V. light sources may also be used but they are less common and will not be included in this review.

The frequencies of a wide variety of radiation sources are listed in Table 3. Ionizing radiations are commercially available as cobalt-60 gamma ray sources or as electron accelerators. The former are low intensity sources but have high depth of penetration, while the latter are high intensity sources but have relatively low depth of penetration.

Table 3. Radiation Source Frequencies [11]

Source	Frequency Hz (cycles/sec)
Cosmic rays	$> 10^{22}$
Gamma rays	3×10^{19} to 10^{22}
X-rays	3×10^{16} to 3×10^{19}
Ultraviolet	10^{15} to 3×10^{16}
Visible	5×10^{14} to 10^{15}
Infrared	10^{12} to 5×10^{14}
Microwaves	10^{9} to 10^{12}
Radiofrequency	2×10^{5} to 10^{9}

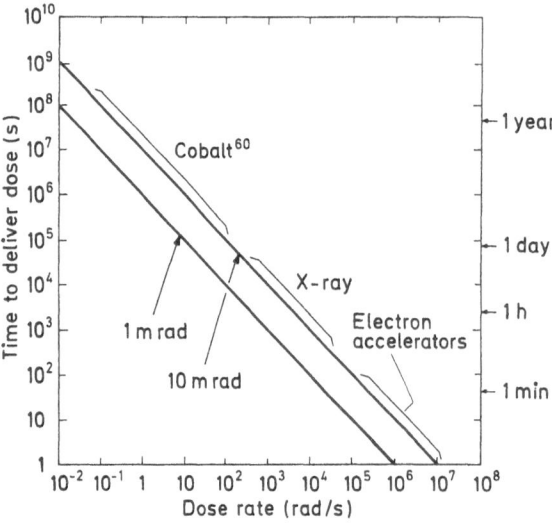

Fig. 2. Effect of dose rate on time required to deliver 1 or 10 Mrad. doses, for radiation of different penetration abilities[12]

Figure 2 compares these two (along with X-ray sources) in the dose range often needed to modify polymer surfaces.

Electric discharges may be classified as "cold" non-equilibrium plasmas or "hot" equilibrium plasmas. Some of the characteristics of these discharges are shown in Table 4.

The energy flow in an electric discharge is described in Table 5 and some of the different types of radicals which may be yielded from vapors or gases added to the discharge are listed in Table 6. Electric discharges may also produce thermal and U.V. radiation, in addition to energetic electrons and ions, atoms and radicals.

Table 4. Types of Electric Discharges [11]

Type	Pressure (torr)	Exit Plasma Character
Silent Discharge (e. g., Ozonizer)	760	"Cold", Non-equilibrium
D. C. Glow Discharge	< 100	"Cold", Non-equilibrium
H. F. Glow Discharge	< 100	"Cold", Non-equilibrium
Arc Discharge-Plasma Torch	760	"Hot", Equilibrium

Table 5. Energy Flow in an Electric Discharge [11]

Electric field + Electron Gas → Energetic Eletrons
Energetic Electrons + Molecular Gas
$\qquad$ → Hot Molecular Gas
$\qquad$ (Excited Molecules
$\qquad$ Atoms and Free Radicals
$\qquad$ Ions)

Atoms + Free Radicals + Molecular Gas $\xrightarrow{\text{Chemical Reaction}}$
$\qquad$ → Reaction Products + Heat of Reaction

Hot Molecular Gas $\xrightarrow[\text{Convection}]{\text{Conduction}}$ Surroundings

Excited Molecules $\xrightarrow{\text{Radiation}}$ Surroundings

$\left.\begin{array}{l}\text{Atoms} \\ \text{Free Radicals} \\ \text{Ions + Elctrons}\end{array}\right\} \xrightarrow{\text{Recombination}}$ Surroundings

Table 6. Some Radicals Produced in Electric Discharges [11]

Diatomic	Triatomic	Polyatomic
CH	NH_2	CH_3
CF	CF_2	C_2H_5
CN	CH_2	CCl_3
CS	PH_2	C_6H_5
NH	CNO	
PH	CNS	
OH		
SH		

3 Surface and Bulk Polymer Modifications, and Synthesis of New Polymeric Systems

There are a number of techniques which can be used to coat substrates with polymers or copolymers (Table 7). Aside from the conventional technique of dipping in a solution of the polymer, other methods involve covalent bonding (grafting) of the polymer to the substrate polymer chains by making use of the free radical sites generated on the surface of the substrate (Fig. 3).

Monomers, which polymerize via a free radical mechanism, can be polymerized on the activated support to produce coatings of various thicknesses and depths of penetration. Ionizing radiation has been extensively used for modifying the surfaces of biomaterials via surface grafting reactions.[4-7]

Hydrophilic, or hydrogel grafts on hydrophobic substrates have been especially studied and a number of reviews of these materials and their biological interactions have been published.[13-20] A bibliography of radiation-grafted biomaterials is presented at the end of the paper.

Ionizing radiation processing, in particular, offers unique advantages for preparing these novel biomaterials. Some of the obvious advantages are that new polymers may be synthesized or existing polymers may be chemically modified by a relatively simple additive-free processing step at room temperature — sometimes with potential for simultaneous sterilization. The doses involved do not generally affect the polymer substrates significantly. Variation of the radiation dose and the grafting solution composition and temperature permits control of the extent of grafting, graft deposition depth, and composition.

The graft copolymer coating process usually produces relatively thick layers, of several microns in depth, which may also penetrate into the polymer substrate matrix. These coatings are usually uniform but they can be rough, and when crystalline poly-

Table 7. Practical Techniques for Depositing Polymer Coatings Onto other Polymeric Substrates

Physical:	Dip-coat substrate in polymer solution and evaporate solvent.
Chemical:	Contact substrate polymer and monomer solution ($\pm$ added polymer) and homopolymerize and/or graft copolymerize using heat and/or catalysts to initiate reaction. Remove solvent and any unreacted monomers.
Chemical, Plasma Discharge or Ionizing Radiation:	Pre-activate polymer substrate (e.g., by peroxidation in air or ozone) using plasma discharge, ionizing radiation or chemical treatment. Then contact with monomer(s) $\pm$ solvents and initiate using heat and/or catalyst to graft copolymerize monomer onto polymer substrate. Remove solvent and any unreacted monomer. ("Indirect Method")
Plasma Discharge or Ionizing Radiation:	Activate polymer substrate and monomer simultaneously using ionizing radiation or plasma gas discharge while monomer and polymer ($\pm$ solvents or other gases) are in direct contact with each other. Remove solvents, gases and any unreacted monomers. ("Direct Method")
U.V.:	Activate polymer substrate and/or monomer with U.V. using mobile or attached photosensitizer, while monomer and polymer ($\pm$ solvents) are in direct contact with each other. Remove solvents and any unreacted monomers.

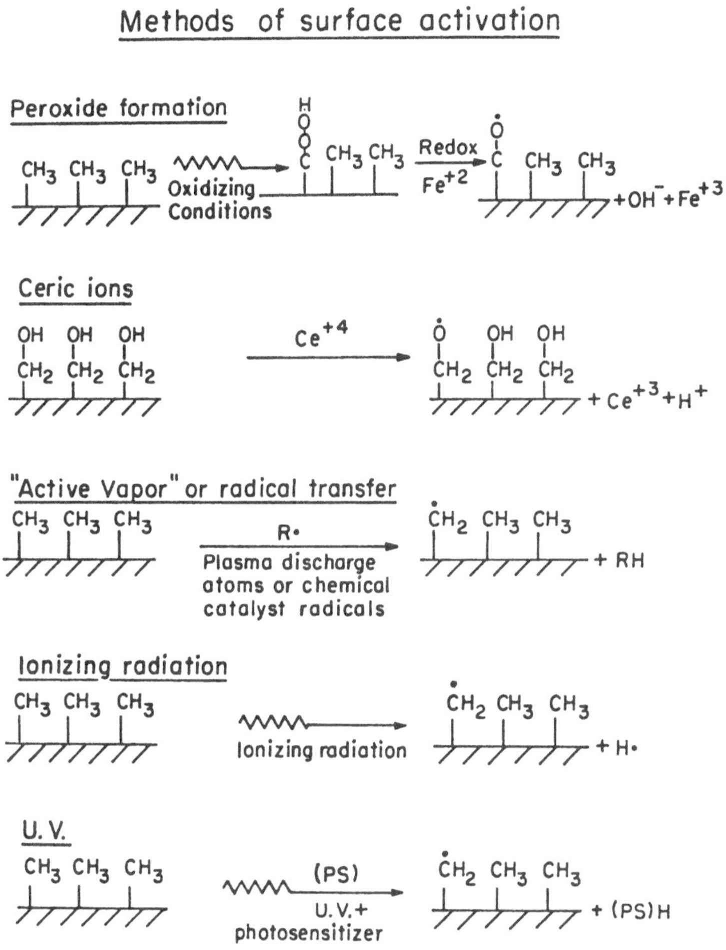

Methods of surface activation

Peroxide formation

$$CH_3\ CH_3\ CH_3 \xrightarrow[\text{Conditions}]{\text{Oxidizing}} \overset{H\text{-}O\text{-}O}{\underset{}{C}}\ CH_3\ CH_3 \xrightarrow[\text{Fe}^{+2}]{\text{Redox}} \overset{O}{\underset{}{C}}\ CH_3\ CH_3 + OH^- + Fe^{+3}$$

Ceric ions

$$\underset{CH_2}{OH}\ \underset{CH_2}{OH}\ \underset{CH_2}{OH} \xrightarrow{Ce^{+4}} \overset{\dot{O}}{\underset{CH_2}{}}\ \underset{CH_2}{OH}\ \underset{CH_2}{OH} + Ce^{+3} + H^+$$

"Active Vapor" or radical transfer

$$CH_3\ CH_3\ CH_3 \xrightarrow[\substack{\text{Plasma discharge}\\ \text{atoms or chemical}\\ \text{catalyst radicals}}]{R\cdot} \overset{\cdot}{C}H_2\ CH_3\ CH_3 + RH$$

Ionizing radiation

$$CH_3\ CH_3\ CH_3 \xrightarrow{\text{Ionizing radiation}} \overset{\cdot}{C}H_2\ CH_3\ CH_3 + H\cdot$$

U. V.

$$CH_3\ CH_3\ CH_3 \xrightarrow[\substack{\text{U.V.} +\\ \text{photosensitizer}}]{(PS)} \overset{\cdot}{C}H_2\ CH_3\ CH_3 + (PS)H$$

Fig. 3. Examples of techniques and reactions for generating radicals on surfaces. (Note: The precise nature of the radical intermediates formed has not been elucidated in some cases. Representations in this figure show schematically radical species which might be formed.)

mers are radiation grafted, the resultant surface often displays "cobblestone-like" roughness with "bumps" of the order of several microns in area.[1,20] The coatings are normally well bonded to the substrate polymer, although thicker hydrophilic grafts on stiff hydrophobic substrates can de-bond if the graft swells extensively in water.

In addition to radiation graft polymerization to modify the surfaces of substrate polymers, one may also use radiation to polymerize pure monomer(s) in solution, or as an emulsion, or in the solid state (e.g., below T_G). One may also radiation crosslink polymers in a solution or in the swollen state, or in the solid state.

Figure 4 and Table 8 summarize the different radiation processing systems which have been studied. Table 9 presents a list of the important considerations and emphases seen in the varied publications in this field. Note that many of the possible syntheses or

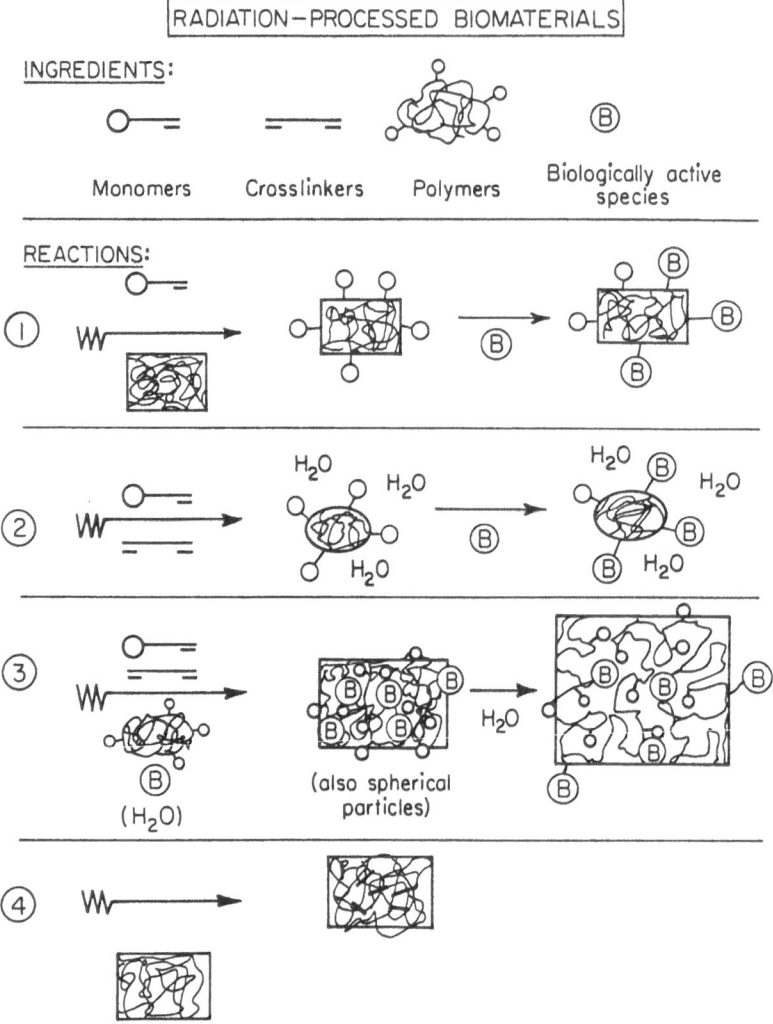

Fig. 4. Major radiation processes used to produce new or modified polymeric biomaterials [6]

modifications include incorporation of biologically functional molecules into the polymeric product. (see below.)

The plasma or glow discharge deposited coatings have different characteristics from radiation grafted surfaces. Their structure and composition depend importantly on the plasma conditions (e. g., gas pressure and flow rate, continuous or pulsed discharge, voltage and energy level, etc.). Their compositional makeup may be broad and contain a variety of chemical groupings, as opposed to the rather well-defined radiation graft polymer. In addition, the coatings may not be deposited uniformly on the substrate surface since the plasma characteristics can vary significantly over very short distances within the reactor.

Table 8. Radiation processing systems [6]

1. Graft Copolymerization of Monomers and Polymers
   a. Surface grafting or surface plus bulk grafting
   b. Immobilization of biologically active species on the graft copolymer
2. Radiation Polymerization (in the Liquid and Solid State, above and below T_G)
   a. Monomers only (includes crosslinkers)
   b. Monomers plus polymers
   c. Monomers plus biologically active species (immobilized during or after polymerization)
   d. Monomers plus polymers plus biologically active species (immobilized during or after polymerization)
3. Radiation Crosslinking
   a. Hydrophobic polymers
   b. Hydrophilic polymers in aqueous solution
   c. Hydrophilic polymers in aqueous solution plus biologically active species (immobilized during or after crosslinking)

Table 9. Important chemical and Biological Considerations in Radiation Processing of Biomaterials [6]

A. Selection and relative amounts of hydrophilic (uncharged, anionic, or cationic) and/or hydrophobic monomers, polymers
B. Processing conditions and synthesis mechanisms, including simultaneous or subsequent immobilization of a biologically active species
C. Physico-chemical characterization of surface and bulk properties
D. *In vitro* interactions with proteins, cells, enzyme substrates, etc.
E. *In vivo* tissue responses (animal model; technique)
F. *In vivo* blood responses (animal model; technique)
G. *In vitro* or *in vivo* release of biologically active species
H. Clinical applications — diagnostics or therapy

Two types of reactions between a gas plasma and a polymer surface can be distinguished. The plasma may react directly with the polymer, e. g., to oxidize it in an oxygen discharge, or the plasma may create polymer radicals by chain transfer reactions and these radicals can then initiate other reactions with the gases, such as graft polymerization. As in the case of ionizing radiation, the predominant reaction will depend on the composition of both the gases and the polymer substrate, and also possibly on the discharge conditions.

The gaseous compounds usefull for plasma deposition may be common gases and saturated or aromatic organic compounds as well as typical vinyl type monomers. The coatings are usually very thin — perhaps only a monomolecular layer — and they are presumed in some cases to be highly crosslinked. One would expect these coatings to remain well-adhered to the substrate polymer under normal conditions. The plasma discharge process only affects the polymer surface, and, conversely, the surface must "see" the plasma to be affected. Biomolecules may be immobilized onto reactable polymer-surface sites after plasma treatment, as in radiation grafting. (See Fig. 4, reaction 1.) Table 10 presents some examples of plasma-treated polymer surfaces prepared for biomedical applications. Additional references on gas discharge-treated biomaterials are cited in the bibliography.

Table 10. Some Examples of Plasma Discharge-Treated Polymer Surfaces Prepared for Biomedical Applications

Gases (or Monomers)	Polymers	Applications	Ref.
NH_3 (or $N_2 + H_2$)	Polypropylene (PP) Poly(Vinyl-chloride) (PVC) Polytetrafluoroethylene (PTFE) Polycarbonate (PC) Polyurethane (PU) Poly(methyl methacrylate) (PMMA)	Heparin bonding for improved blood compatibility	[21]
Hexamethyldisiloxane (HMDS) $C_2H_4 + N_2$ Allene $+ N_2 + H_2O$	Poly(ethylene terephthalate) (PET) Silastic (SR) Polysulfone (PS)	Improved blood compatibility (in some cases)	[22]
Hexamethyl- and Octamethylcyclotetrasiloxane	PP	Improved membrane for blood oxygenator	[23]
C_2H_4, allene, styrene, acrylonitrile, C_2F_4, C_2H_3F, C_2F_3Cl, C_2H_3Cl	Polystyrene (PSt) SR	Improved tissue compatibility	[24]
C_2H_4, C_2F_3Cl, styrene	SR	Improved tissue compatibility	[25]
$C_2H_2 + N_2 + H_2O$	PMMA	Modify corneal contact lens wettability by proteins	[26]
C_2H_4, Ar	PP, PET, PVC, SR, Poly(methyl acrylate) (PMA)	Reduce leaching of small molecules from polymer into body	[27]
C_2F_4, Et_3SiH, pyridine	PVC (DOP plasticized)	Reduce leaching of DQP into blood	[28]
C_2H_4, C_2F_4, C_2H_6, Ar	Poly(2-hydroxyethyl methacrylate) (Poly(HEMA)) Poly(HEMA-MA)	Control of pilocarpine release rate from hydrogel	[29]
C_2H_4, C_2F_4, C_2H_6, Ar	SR	Reduce progesterone release rate from SR	[30]

4 Immobilization of Biomolecules

A wide variety of biologically active species may be incorporated into or onto radiation or plasma processed polymeric biomaterials (Fig. 4) for a wide variety of uses (Tables 11–13). Such species are usually present in aqueous solutions, or if not, at least function optimally in aqueous systems. Therefore, most of the monomers and

Table 11. Biologically Active Species which may be Immobilized within or on Radiation-Processed Polymeric Biomaterials

Enzymes	Contraceptives
Antibodies	Anticancer agents
Antigens	Drug antagonists
Anti-thrombogenic agents	Other drugs, in general
Antibiotics	Sugars and polysaccharides
Antibacterial agents	Cells

Table 12. Some Examples of uses of Immobilized Bio-molecule-Polymer Systems

Improved biocompatibility
Drug delivery
Cell "finders" and "markers"
  (via antibody-antigen binding)
Diagnostic kits
Enzyme reactors (including artifical organs)
Biomedical sensors or electrodes

Table 13. Biomedical Applications of Immobilized Enzymes [19]

Immobilized Enzyme(s)	Application
Brinolase	Non-thrombogenic surface
Urokinase	Non-thrombogenic surface
Streptokinase	Non-thrombogenic surface
Asparaginase, Glutaminase	Leukemia treatment
Carbonic Anhydrase, Catalase	Membrane oxygenator
Urease	Artificial kidney
Glucose Oxidase	Glucose sensor — artificial pancreas
Microsomal enzymes	Artificial liver
Alcohol Oxidase	Blood alcohol electrode
LNase, RNase	Removal of airborn infections

polymers involved in immobilizing such species have dominant hydrophilic character. One may immobilize the biologically active molecules or cells by four major techniques: (1) physical entrapment; (2) electrostatic attraction; (3) chemical bonding; and (4) physical adsorption plus chemical crosslinking.

Physical entrapment is the simplest, since it only involves radiation polymerization of monomers $\pm$ polymers, or radiation crosslinking of polymers. However, it is very important that either a significant pore structure exist in the final product or that it be in a finely divided form, so as to provide access for other biomolecules to reach the immobilized biomolecule, or vice versa. (The word "immobilization" refers in a temporal sense to the suddenly lowered mobility of the biological species; it may never leach out or it may gradually dissolve into the surrounding medium. In either case, it is considered to have been "immobilized".)

Electrostatic attraction is like affinity chromatography. It may be used to immobilize polyanionic biomolecules as heparin onto cationic sites, or to bind an antigen onto an already immobilized antibody, or a substrate molecule onto an immobilized enzyme. Specific biological binding requires previous immobilization of the specific binding site. Hydrophobic binding is a special case of electrostatic attraction as an immobilization mechanism.

Chemical bonding involves specialized chemical reactions on specific backbone groups (usually $-OH$ or $-CO_2H$) in order to activate these sites so they can form primary bonds with the species to be immobilized (usually via $-NH_2$ groups on such species). This is the most complex technique of all and may involve several steps. There has been extensive work using radiation processing plus chemical immobilization techniques. Radiation grafted hydrogels or radiation polymerized emulsions of HEMA or MAAc and their copolymers have been most used for subsequent immobilization of biomolecules. Rembaum and coworkers have pioneered in this area, immobilizing antibodies on radiation-polymerized polymeric emulsions.[31-35] In addition, it should be noted that the radiation itself may lead to chemical bonding between the biomaterials and the polymer matrix. Thus, a method in which a physically entrapped biomolecule-monomer/polymer system is irradiated may, in fact, also chemically immobilize the biomolecule.

Plasma treatments are useful only for surface immobilization of biomolecules. This technique can be much less "precise" than radiation-grafted surface immobilization due to the larger possible number of chemical groups produced on a plasma treated surface.

Surface adsorption followed by chemical crosslinking of the unadsorbed biomolecule is another method for immobilization. Radiation processing is not a necessary step in this technique.

5 Conclusion

The application of radiation and gas discharge processing to synthesize or modify polymeric materials for medical applications continues to increase in interest and diversity around the world. Incorporation of biologically active molecules is an important additional processing step which can open up many new and exciting medical applications for diagnosis and therapy.

Acknowledgment: The author would like to thank Liz Zielie for typing the manuscript and Hedi Nurk for preparing the figures. He also gratefully acknowledges the support of the NHLBI, Program Project grant HL-22163-05.

6 References

A. Cited in Text

1. Hoffman, A. S.: Synthetic polymer biomaterials — a review, in: IUPAC — Macromolecules (ed.) H. Benoit and P. Rempp, p. 321, London, Pergamon Press 1982
2. Hoffman, A. S.: J. Appl. Polym. Sci., Appl. Polym. Symp. *31*, 313 (1977)

3. Hoffman, A. S.: J. Biomed. Matls. Res. Symp. *5(1)*, 77 (1974)

4. Hoffman, A. S.: Rad. Phys. Chem. *9*, 207 (1977)

5. Machi, S., Ishigaki, T.: Genshiryoku Kogyo *24(5)*, 45 (1978)

6. a) Hoffman, A. S.: Rad. Chem. (Japan) *16(31)*, 12 (1981)
   b) Hoffman, A. S.: Rad. Phys. Chem. *18(1)*, 323 (1981)

7. Hoffman, A. S. et al.: Rad. Phys. Chem. Vol. 22, Nos. $^1/_2$, pp. 267–283, 1983

8. Hollahan, J. R., Bell, A. T.: Techniques and Applications of Plasma Chemistry, New York, Wiley-Interscience 1974

9. Shen, M., Bell, A. T. (eds.): Plasma Polymerization, ACS Symposium Series 108, Washington, D.C., American Chemical Society 1979

10. Yasuda, H., Gaziki, M.: Biomaterials 2, 68 (1982)

11. Hoffman, A. S.: Surface modifications of polymers for biomedical applications, in: Science and Technology of Polymer Processing (eds.) N. P. Suh and N. H. Sung, p. 200, Cambridge, Mass., MIT Press 1979

12. Hoffman, A. S.: Atomic Energy Rev. 9, 2,347 (1971)

13. Wichterle, O., Lim, D.: Nature *185*, 117 (1960)

14. Barvic, M., Kliment, K., Zavadil, M.: J. Biomed. Matls. Res. *1*, 313 (1967)

15. Sprincl, L., Kopecek, J., Lim, D.: J. Biomed. Matls. Res. *4*, 447 (1971)

16. Abrahams, R. A., Ronel, S. H.: ACS Polym. Preprints *16*, 535 (1975)

17. Hoffman, A. S., Ratner, B. D.: ACS Polym. Preprints *16(2)*, 272 (1975)

18. Hoffman, A. S.: Hydrogels — a broad class of biomaterials, in: Polymers in Medicine and Surgery (eds.) R. L. Kronenthal, Z. Oser and E. Martin, New Jersey, Plenum Press 1975

19. Ratner, B. D., Hoffman, A. S.: Synthetic hydrogels for biomedical applications, in: Hydrogels for Medical and Related Applications, ACS Symposium Series 31 (ed.) J. D. Andrade, p. 1, Washington, D.C., American Chemical Society 1976

20. Ratner, B. D., Hoffman, A. S.: Surface grafted polymers for biomedical applications, in: Synthetic Biomedical Polymers: Concepts and Applications (eds.) W. J. Robinson and M. Szcher, p. 133, Westport, Connecticut, Technomic Publishing Co. 1980

21. Hollahan, J. R., Stafford, B. B., Falb, R. D.: J. Appl. Polym. Sci. *13*, 807 (1969)

22. Yasuda, M., Bumgarner, M. O., Mason, L. G.: Improvement of blood compatibility of membranes by discharge polymerization, in: Permeability of Plastic Films and Coatings to Gases, Vapors and Liquids, Polymer Science and Technology, Volume 6 (ed.) H. P. Hopfenberg, p. 453, New York, Plenum Press 1974

23. Chawla, A. S.: Art. Org. *3(1)*, 92 (1979)

24. Hahn, A. W. et al.: NBS Spec. Publ. *415*, 13 (1975)

25. Nichols, M. F. et al.: J. Biomed. Matls. Res. *13*, 299 (1979)

26. Yasuda, H. et al.: J. Biomed. Matls. Res. *9*, 629 (1975)

27. Chang, F. Y., M. Shen, Bell, A. T.: J. Appl. Polym. Sci. *17*, 2915 (1973)

28. Hatada, K., Kobayashi, H., Asai, M.: Org. Coat. and Appl. Polym. Sci. Proc. *47*, 391 (1982)

29. Colter, K. D., Bell, A. T., Shen, M.: Biomatls., Med. Dev., Art. Org. *5(1)*, 1 (1977)

30. Colter, K. D., Bell, A. T., Shen, M.: Biomatls., Med. Dev., Art. Org. *5(1)*, 13 (1977)

31. Yen, S. P. S. et al.: ACS Polym. Preprints *16(1)*, 181 (1975)

32. Rembaum, A. et al.: Macromolecules 9, 328 (1976)

33. Rembaum, A., Yen, S. P. S., Volksen, W.: Chem. Tech. *8(3)*, 182 (1978)

34. Rembaum, A.: ACS Polym. Preprints *20*, 354 (1979)

35. Rembaum, A., Yen, S. P. S., Molday, R. S.: J. Macromol. Sci.-Chem. *A13*, 603 (1979)

B. *Additional References on Radiation-Grafted Hydrogels for Biomedical Applications*

36. Yasuda, H., Refojo, M. F.: J. Polym. Sci. *A2*, 5093 (1964)

37. Laizier, J., Wajs, G.: Resines silicones hydrophiles obtenues par greffage radiochimique, in: Large Radiation Sources for Industrial Process, p. 205, Vienna, IAEA 1969

38. Hoffman, A. S., Kraft, W. G.: ACS Polym. Preprints *13(2)*, 723 (1972)

39. Hoffman, A. S., Harris, C.: ACS Polym. Preprints *13(2)*, 740 (1972)

40. Lee, H. B., Shim, H. S., Andrade, J. D.: ACS Polym. Preprints *13*, 729 (1972)

41. Ratner, B. D., Hoffman, A. S.: ACS Org. Coat. Plast. Chem. Preprints 33, 286 (1973)

42. Kearney, J. J., Amara, J. J., McDevitt, M. B.: ACS Polymer. Preprints *14*, 346 (1973)

43. Ratner, B. D., Hoffman, A. S.: J. Appl. Polym. Sci. *18*, 3183 (1974)
44. Hoffman, A. S. et al.: Preparation and application of radiation-grafted hydrogels as biomaterials, in: The Permeability of Plastic Films and Coatings to Gases, Vapors and Liquids (ed.) H. P. Hopfenberg, p. 441, New York, Plenum Press 1975; also ACS Org. Coat. Plast. Chem. Abstracts *34(1)*, 568 (1974)
45. Meaburn, G. M. et al.: Radiation-induced grafting of hydrogels on silicone rubber surfaces, in: Abstracts of Papers, Fifth International Congress on Radiation Research, p. 200, Seattle, Wash. 1974
46. Ratner, B. D., Hoffman, A. S.: ACS Org. Coat. Plast. Chem. Preprints *33(2)*, 386 (1973); also Polymer Sci. Tech. *Series No. 7*, 159 (1975)
47. Ronel, S. H., Abrahams, R. A.: Soc. Plastics Eng. Tech. Papers *21*, 570 (1975)
48. Sasaki, T., Ratner, B. D., Hoffman, A. S.: ACS Polym. Preprints *16(2)*, 435 (1975); also ACS Symposium Series *No. 31*, 283 (1976)
49. Ratner, B. D., Balisky, T., Hoffman, A. S.: J. Bioeng. *1*, 115 (1977)
50. Greer, R. T., Vale, B. H., Knoll, R. L.: Scanning Electron Microscopy *1*, 633 (1978)
51. Hoffman, A. S., Ratner, B. D.: ACS Polym. Preprints *20*, 423 (1979)
52. Hoffman, A. S., Ratner, B. D.: Rad. Phys. Chem. *14*, 831 (1979)
53. Chapiro, A. et al.: Rad. Phys. Chem. *15*, 423 (1980)
54. Chapiro, A. et al.: preprint, courtesy of the authors
55. Peppas, N. A., Gehr, T. W. B.: Trans. Amer. Soc. Artif. Int. Organs *24*, 404 (1978)
56. Peppas, N. A., Gehr, T. W. B.: J. Appl. Polym. Sci. *24*, 2159 (1979)
57. Kamel, I. L.: Rad. Phys. Chem. 9, 711 (1977)
58. Khaw, B., Ratner, B. D., Hoffman, A. S.: ACS Polym. Preprints *16(2)*, 446 (1975); also ACS Symposium Series *No. 31*, 295 (1976)
59. Ratner, B. D. et al.: J. Polym. Sci. *66*, 363 (1979)
60. Ratner, B. D., Hoffman, A. S., Whiffen, J. D.: Biomatls. Med. Dev. and Artif. Orgs. *3(1)*, 115 (1975)
61. Ratner, B. D., Hoffman, A. S., Whiffen, J. D.: J. Bioeng. *2*, 313 (1978)
62. Hanson, S. R. et al.: Ann. Biomed. Eng. *7*, 357 (1979)
63. Hanson, S. R. et al.: Evaluation of artificial surfaces using a baboon arteriovenous shunt model, in Proc. of 1st World Biomaterials Congress, Baden, Austria; London, John Wiley and Sons 1980
64. Hanson, S. R. et al.: Clin. Med. *95*, 289 (1980)
65. Hoffman, A. S., Horbett, T. A., Ratner, B. D.: Ann. N.Y. Acad. Sci. *283*, 372 (1977)
66. Thon, M. S., Andrade, J. D.: J. Biomed. Matls. Res. *7*, 509 (1973)
67. Hoffman, A. S. et al.: Estimation of surface energy components and their potential relevance to biological interactions, in Proceedings of NSF Workshop on Interfacial Phenomena (ed.) J. C. Berg, p. 445, Seattle, Washington, University of Washington 1979
68. Ko, Y. C., Ratner, B. D., Hoffman, A. S.: J. Coll. Interfac. Sci. *82*, 25 (1981)
69. Ratner, B. D.: J. Biomed. Matls, Res. *14*, 665 (1980)
70. Ratner, B. D., Hoffman, A. S.: Surface characterization of hydrophilic-hydrophobic copolymer model systems. I. A preliminary study, in Adhesion and Adsorption of Polymers, Part B (ed.) L. H. Lee, p. 691, New York, Plenum Press 1980
71. Ratner, B. D. et al.: J. Appl. Polym. Sci. *22*, 643 (1977)
72. Ratner, B. D., Hoffman, A. S.: ACS Org. Coat. Plast. Chem. Preprints *40*, 714 (1979)
73. Weathersby, P. K. et al.: J. Bioeng. *1(4)*, 381 (1977)
74. Miller, M. L. et al.: J. Appl. Polym. Sci. *14*, 257 (1970)
75. Ziegler, T. F., Miller, M. L.: J. Biomed. Matls. Res. *4*, 259 (1970)
76. Lundell, E. O. et al.: ACS Polym. Preprints *16*, 541 (1975)
77. Horbett, T. A., Hoffman, A. S.: ACS Advances in Chemistry Series *No. 145*, Applied Chemistry at Protein Interfaces, p. 230 (1975)
78. Horbett, T. A., Weathersby, P. K., Hoffman, A. S.: Thrombosis Research *12*, 319 (1978)
79. Horbett, T. A.: The kinetics of adsorption of plasma proteins to a series of hydrophilic-hydrophobic copolymers, in: Adhesion and Adsorption of Polymers, Part B (ed.) L. H. Lee, p. 677, New York, Plenum Publishing Corp. 1980
80. Horbett, T. A.: Surface enrichment of plasma proteins, in: Proceedings of 1st World Biomaterials Congress (Baden, Austria), London, John Wiley and Sons 1980

81. Weathersby, P. K., Horbett, T. A., Hoffman, A. S.: Trans. Amer. Soc. Artif. Int. Organs 22, 242 (1976)
82. Weathersby, P. K., Horbett, T. A., Hoffman, A. S.: J. Bioeng. 1(4), 395 (1977)
83. Harker, L. A., Hanson, S. R., Hoffman, A. S.: Ann. N.Y. Acad. Sci. 283, 317 (1977)
84. Ratner, B. D. et al.: J. Biomed. Matls. Res. 9, 407 (1975)
85. Sprincl, L., Vacik, J., Kopecek, J.: J. Biomed. Matls. Res. 7, 123 (1973)
86. Gilding, D. K. et al.: Trans. Amer. Soc. Artif. Int. Organs 24, 411 (1978)
87. Kronik, P. L., Rembaum, A.: ACS Polym. Preprints 16, 157 (1975)
88. Muzykewicz, K. J. et al.: J. Biomed. Matls. Res. 9, 487 (1975)
89. Ihlenfeld, J. V. et al.: Trans. Amer. Soc. Artif. Int. Organs 24, 627 (1978)
90. Hasegawa, T. et al.: Surgery 74, 696 (1973)
91. Hoffman, A. S. et al.: Thrombotic events on grafted polyacrylamide-Silastic surfaces as studied in a baboon, in: Morphology and Interactions of Biomaterials (eds.) S. L. Cooper, A. S. Hoffman, N. A. Peppas and B. D. Ratner, p. 59, ACS Advances in Chemistry Series No. 199, York, PA, American Chemical Society 1982
92. Hoffman, A. S. et al.: Artif. Organs 5(Suppl), 540 (1981)
93. Fischer, J. P. et al.: J. Polym. Sci., Polym. Symp No. 66, 443 (1979)
94. Fischer, J. P. et al.: Proc. of 1st World Biomaterials Congress (Baden, Austria), London, John Wiley and Sons 1980
95. Jansen, B., Ellinghorst, G.: J. Polym. Sci., Polym. Symp. No. 66, 465 (1979)
96. Rösinger, S., Fischer, J. P., Fuhge, P.: "Development of Materials with Blood Compatibility by Radiation Processing," in IAEA Conference Proceedings, Genoble, September 28-October 2 1981, p. 323, Vienna, IAEA 1981
97. Gaussens, G.: Les nouveaux biomateriaux obtenus par l'action des rayonnements ionisants, in: IAEA Conference Proceedings, Grenoble, September 28-October 1 1981, p. 343, Vienna, IAEA 1981

C. *Additional References on Radiation Crosslinked Polymers for Biomedical Applications*

 98. Bray, J. C., Merrill, E. W.: J. Biomed. Matls. Res. 7, 433 (1973)
 99. Bray, J. C., Merrill, E. W.: J. Appl. Polym. Sci. 17, 3779 (1973)
100. Peppas, N. A., Merrill, E. W.: J. Biomed. Matls. Res. 11, 423 (1977)
101. Ikada, Y., Mita, T.: Rad. Phys. Chem. 9, 633 (1977)
102. Kalal, J., Bednar, B., Houska, M.: ACS Polym. Preprints 16, 363 (1975)
103. Vlasova, N. K. et al.: Stomatologiya (Moscow) 50, 33 (1971)
104. Weathersby, P. K., Kolobow, T., Stool, E. W.: J. Biomed. Matls. Res. 9, 561 (1975)
105. Chawla, A. S.: J. Bioeng. 2, 231 (1978)
106. Gifford, G. H., Jr., Merrill, E. W., Morgan, M. S.: J. Biomed. Matls. Res. 10, 857 (1976)
107. du Plessis, T. A., Grobbelaar, C. J., Marais, F.: Rad. Phys. Chem. 9, 647 (1977)
108. Grobbelaar, C. J., du Plessis, T. A., Marais, F.: J. Bone Joint Surg. (London) 60B, 370 (1978)
109. Shimizu, Y. et al.: Biomatls. Med. Dev. Artif. Organs 6, 375 (1978)
110. Al-Saden, A. A. et al.: Internatl. J. Pharmaceutics 5, 317 (1980)
111. Gaussens, G., Nicaise, M., Tran, K. O.: Protheses articulaires a base de polymeres ameliorés sous rayonnements ionisants, in: IAEA Conference Proceedings, Grenoble, September 28-October 2 1981, p. 373, Vienna, IAEA 1981

D. *Additional References on Radiation-Processed Biomaterials Cäntaining Immobilized Biomolecules*

112. Dobo, J.: Acta Chim. Acad. Sci. Hung. 63, 453 (1970)
113. Maeda, H., Yamauchi, A., Suzuki, H.: Biochim. Biophys. Acta 315, 18 (1973)
114. Maeda, H., Suzuki, H., Yamauchi, A.: Biotech. Bioeng. 15, 827 (1973)
115. Maeda, H. et al.: Biotech. Bioeng. 17, 119 (1975)
116. Maeda, H., Suzuki, H.: Proc. Biochem. 12(6), 9 (1977)
117. Kawashima, K., Umeda, K.: Biotech. Bioeng. 16, 609 (1974)
118. Kawashima, K., Umeda, K.: Agric. Biol. Chem. 40(6), 1143 (1976)
119. Kawashima, K., Umeda, K.: Agric. Biol. Chem. 40(6), 1151 (1976)
120. Kumakura, M. et al.: J. Solid Phase Biochem. 2, 279 (1977)
121. Kaetsu, I., Kumakura, M., Yoshida, M.: Biotech. Bioeng. 21, 847 (1979)

122. Kaetsu, I., Kumakura, M., Yoshida, M.: Biotech. Bioeng. *21*, 863 (1979)
123. Kaetsu, I. et al.: Rad. Phys. Chem. *14*, 595 (1979)
124. Kaetsu, I. et al.: J. Biomed. Matls. Res. *14*, 199 (1980)
125. Yoshida, M., Kumakura, M., Kaetsu, I.: Polymer *20*, 3 (1979)
126. Yoshida, M., Kumakura, M., Kaetsu, I.: Polymer *20*, 9 (1979)
127. Yoshida, M., Kumakura, M., Kaetsu, I.: Polymer J. *11*, 915 (1979)
128. Kumakura, M., Yoshida, M., Kaetsu, I.: Appl. Env. Microbiol. *37*, 310 (1979)
129. Kumakura, M., Yoshida, M., Kaetsu, I.: Biotech. Bioeng. *21*, 679 (1979)
130. Yoshida, M., Kumakura, M., Kaetsu, I.: Polymer *19*, 1375 (1978)
131. Yoshida, M., Kumakura, M., Kaetsu, I.: Polymer *19*, 1379 (1978)
132. Yoshida, M., Kumakura, M., Kaetsu, I.: J. Pharmaceut. Sci. *68*, 628 (1979)
133. Yoshida, M., Kumakura, M., Kaetsu, I.: Polymer J. *11*, 775 (1979)
134. Yoshida, M., Kumakura, M., Kaetsu, I.: J. Pharmaceut. Sci. *68*, 860 (1979)
135. Kaetsu, I., Yoshida, M., Yamada, A.: J. Biomed. Matls. Res. *14*, 185 (1980)
136. Kaetsu, I. et al.: Biomaterials *1*, 17 (1980)
137. Leininger, R. I. et al.: Trans. Amer. Soc. Artif. Int. Organs *12*, 155 (1966)
138. Leininger, R. I. et al.: Science *152*, 1625 (1966)
139. Falb, R. D. et al.: J. Biomed. Matls. Res. *1*, 239 (1967)
140. Leininger, R. I., Falb, R. D., Grode, G. A.: Ann. N.Y. Acad. Sci. *146*, 11 (1968)
141. Rembaum, A., Singer, S., Keyzer, H.: Polym. Letters *7*, 395 (1969); also NASA Tech. Brief, No. 69-10299, August 1969
142. Chawla, A. S., Chang, T. M. S.: ACS Polym. Preprints *14*, 379 (1973); also Biomatls. Med. Dev. Artif. Orgs. *2*, 157 (1972)
143. Hoffman, A. S. et al.: Trans. Amer. Soc. Artif. Int. Orgs. *18*, 10 (1972)
144. Hoffman, A. S., Schmer, G.: New approaches to non-thrombogenic materials, in: Current Topics in Coagulation (ed.) G. Schmer, p. 201, New York, Academic Press 1973
145. Hoffman, A. S., Schmer, G.: Paroi Arterielle — Arterial Wall *1*, 95 (1973)
146. Mate, T. P. et al.: pH effects in the covalent coupling of small molecules and proteins to poly (HEMA-methacrylic acid) hydrogels, in: Enzyme Engineering (eds.) Pye and Wingard, p. 137, New York, Plenum Press 1975
147. Smiley, K. L., Boundy, J. A.: Presented at American Chemical Society Meeting, Chicago, 1973
148. Venkataraman, S., Horbett, T. A., Hoffman, A. S.: ACS Polym. Preprints *16(2)*, 197 (1975); also J. Biomed. Matls. Res. *11*, 111 (1977)
149. Venkataraman, S., Horbett, T. A., Hoffman, A. S.: J. Mol. Catalysis *2*, 273 (1977)
150. Barker, H. et al.: A radiation grafting technique for the immobilization of enzymes and the heterogenizing of catalytically active homogeneous metal complexes, in: Proceedings of the VIth Internatl. Congr. Catalysis, p. 551, London, The Chemical Society 1977
151. Liddy, M. J., Garnett, J. L., Kenyon, R. S.: J. Polym. Sci., *Symp. No. 49*, 109 (1975)
152. Molday, R. S. et al.: Nature *249*, 81 (1974)
153. Molday, R. S. et al.: J. Cell. Biol. *64*, 75 (1975)
154. Gordon, I. L. et al.: Cell Immunol. 28, 307 (1977)
155. Catt, K., Niall, H. D., Tregear, G. W.: Biochem. J. *100*, 31c (1966)
156. Catt, K., Niall, H. D., Tregear, G. W.: J. Lab. Clin. Med. *70*, 820 (1967)
157. Catt, K., Niall, H. D., Tregear, G. W.: J. Exp. Biol. Med. Sci. *45*, 703 (1967)
158. Catt, K., Niall, H. D., Tregear, G. W.: J. Clin. Endocrin. Metabolism *28*, 121 (1968)
159. Tregear, G. W.: in Solid Phase Proteins — Their Preparation, Properties and Applications (ed.) A. S. Hoffman, Proceedings of a Conference at Battelle-Seattle Research Center, September 1971
160. Kobayashi, M., Kaetsu, I.: Physical immobilization of biofunctional substance by the use of radiation polymerization, in: IAEA Conference Proceedings, Grenoble, September 28-October 2, 1981, p. 353, Vienna, IAEA 1981

E.   *Additional References on Gas Discharge Polymer Treatments for Biomedical Applications*

161. Scott, H., Hillman, E. E.: Active-vapor grafting of hydrogels in medical prostheses, Contract No. NIH-HHL1-71-2017, National Heart and Lung Institute, National Institutes of Health, Bethesda, Maryland, Annual Report (February 1, 1973), PB-221-846
162. Smith, L. et al.: ACS Polym. Preprints *16(2)*, 186 (1975)

163. Yasuda, H. et al.: J. Biomed. Matls. Res. *9*, 629 (1975)
164. Yasuda, H., Bumgarner, M. O., Mason, R. G.: Biomatls., Med. Dev., Artif. Int. Orgs. *4*, 307 (1976)
165. Chawla, A. S.: Trans. Amer. Soc. Artif. Int. Orgs. *25*, 287 (1979)
166. Nichols, M. F. et al.: J. Biomed. Matls. Res. *13*, 299 (1979)
167. Hahn, A. W. et al.: Biomed. Scient. Instrumentation *15*, 7 (1979)
168. Yasuda, H.: A study of electrodeless glow discharge as a means of modifying the surface of polymers Annual Reports No. NIH-NO1-HV-3-2913-1 through NIH-NO1-HV-3-2913-7 (1972–1979)
169. Cannon, J. G. et al.: J. Biomed. Matls. Res. *14*, 279 (1980)
170. Özdural, A. R. et al.: Am. Soc. Artif. Int. Organs J. *3*, 116 (1980)
171. Hahn, A. W. et al.: Biomed. Sci. Instrumentation *17*, 109 (1981)
172. Nichols, M. F. et al.: Biomaterials *2*, 161 (1981)
173. Sadhir, R. K. et al.: Biomaterials *2*, 239 (1981)
174. Yasuda, H.: J. Polymer Sci., Macromol. Rev. *16*, 199 (1981)
175. Hahn, A. W. et al.: Org. Coat. and Appl. Polym. Sci. Proc. *47*, 386 (1982)
176. Matsuzawa, Y., Yasuda, H.: Org. Coat. and Appl. Polym. Sci. Proc. *47*, 397 (1982)

K. Dušek (Editor)
Received May 31, 1983

Author Index Volumes 1–57

Subject Index

B.-O. Küppers

Molecular Theory of Evolution

Outline of a Physico-Chemical Theory of the Origin of Life

Translated from the German by P. Woolley
1983. 76 figures. IX, 321 pages. ISBN 3-540-12080-7

Contents: Introduction. – The Molecular Basis of Biological Information: Definition of Living Systems. Structure and Function of Biological Macromolecules. The Information Problem. – Principles of Molecular Selection and Evolution: A Model System for Molecular Self-Organization. Deterministic Theory of Selection. Stochastic Theory of Selection. – The Transition from the Non-Living to the Living: The Information Threshold. Self-Organization in Macromolecular Networks. Information-Integrating Mechanisms. The Origin of the Genetic Code. The Evolution of Hypercycles. – Model and Reality: Systems Under Idealized Boundary Conditions. Evolution in the Test-Tube. Conclusions: The Logic of the Origin of Life. – Mathematical Appendices. – Bibliography. – Index.

The subject of this book is the physico-chemical theory of the origin of life. Although this theory is still in statu nascendi, it has been developed to the point where a coherent presentation is desirable and possible. The theory arises from the application of the principles and methods of physics and chemistry to modern biology. In essence, it provides a physical extension of the classical Darwinian concept of evolution to the molecular level and shows how this new concept can be applied to the problem of the origin of biological information.

This work investigates the problem of how life arose and distinguishes at least three phases: a phase of chemical evolution, a phase of molecular self-organization and a phase of biological evolution. The actual transition from non-living matter to living is clearly to be attributed to the phase of molecular self-organization, so this is the phase with which the monograph is concerned.

The monograph is intended as an introductory text for students of physics, chemistry or biology who are interested in obtaining an overall view of the present state of the theory of the origin of life. This interdisciplinary aim has necessitated a choice of material based in the lowest common denominator of physicists and biologists. In particular the more mathematical and biological sections have been at the most elementary level possible (139 ref.).

Springer-Verlag
Berlin
Heidelberg
New York
Tokyo

P. F. Gordon, P. Gregory

Organic Chemistry in Colour

1983. 52 figures, 59 tables. XI, 322 pages
ISBN 3-540-11748-2

Contents: The Development of Dyes. – Classification and Synthesis of Dyes. – Azo Dyes. – Anthraquinone Dyes. – Miscellaneous Dyes. – Application and Fastness Properties of Dyes. – Author Index. – Subject Index.

Organic Chemistry in Colour emphasizes the strong links that exist between dyestuffs and organic chemistry. The most important properties of dyestuffs are discussed in terms of modern organic chemistry, with special emphasis on current molecular orbital theories. Dye synthesis is discussed in the light of modern synthetic methods and, where appropriate, current thinking on mechanistic aspects is considered.

The book therefore provides an ideal forum for those seeking an insight into modern organic chemistry whilst simultaneously seeing its application to an important industrial field. To this end, then, the book should fulfill a dual function both as useful reference for research workers in the field of organic chemistry and dyes, and also as an aid to the advanced chemistry student who would like to see organic chemistry illustrated by practical examples.

Springer-Verlag
Berlin
Heidelberg
New York
Tokyo